Roy V. Hughson   English For Careers

The Language of Chemical Engineering in English

Prentice Hall Regents, Englewood Cliffs, NJ 07632

Drawings by Bernie Case

Roy V. Hughson is Associate Editor of
magazine *Chemical Engineering*

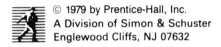 © 1979 by Prentice-Hall, Inc.
A Division of Simon & Schuster
Englewood Cliffs, NJ 07632

Printed in the United States of America

10 9 8 7 6 5 4 3 2

ISBN 0-13-523242-2 01

Prentice-Hall International (UK) Limited, *London*
Prentice-Hall of Australia Pty. Limited, *Sydney*
Prentice-Hall Canada Inc., *Toronto*
Prentice-Hall Hispanoamericana, S.A., *Mexico*
Prentice-Hall of India Private Limited, *New Delhi*
Prentice-Hall of Japan, Inc., *Tokyo*
Simon & Schuster Asia Pte. Ltd., *Singapore*
Editora Prentice-Hall do Brasil, Ltda., *Rio de Janeiro*

1063266-2

TABLE OF CONTENTS

FOREWORD

This book is one of a series called *English for Careers*, intended to introduce students of English to the specialized language of a number of professions and vocations. The career areas covered are those in which English is widely used throughout the world.

Each book in this series serves a dual purpose: to give the English student an introduction to the English terminology of the vocational area in which he or she is interested, and to improve the overall use of this language. This book describes the work of chemical engineers—the tools, instruments, and processes they use, the places in which they work, and some of their contributions to the present and goals for the future.

With respect to learning English as a second language, these books are intended for a student at the high intermediate or advanced level who is acquainted with most of the structural patterns of English. The principal goals of the learner should be mastering specific vocabulary, using the patterns in a normal mixture, and perfecting his or her ability to communicate in this language, especially in the chosen career area.

This book meets these needs. Each unit begins with a glossary in which specialized words and expressions are defined. This is followed by vocabulary practice that tests the student's comprehension of the special terms and gives practice in their use. In the reading that follows, these terms are used within a contextual frame of reference. Each selection is followed by exercises for comprehension, discussion, and review. They give the student the opportunity to use, in a communicative situation, both the vocabulary and the structural patterns that have occurred in the reading. Each unit ends with exercises which provide additional contexts for the terms as well as conversational and/or writing practice. The index of special terms at the end of the book makes it easy to use the specialized vocabulary for review or reference.

Much successful language learning comes from interest and experience which is not fully conscious. In offering this book, it is hoped that the student's interest in chemical engineering will enhance his or her ability to communicate more effectively in English.

ROY V. HUGHSON
New York, New York

UNIT ONE
WHAT CHEMICAL ENGINEERS DO

Special Terms

Chemical Process Industries (CPI): A large group of industries that use chemical and engineering principles to separate or change materials into salable products.

Raw Material: The material that comes into a plant, where it is processed to produce a salable product. Petroleum is the raw material for the manufacture of gasoline. Sulfur is the raw material for the manufacture of sulfuric acid (H_2SO_4).

Unit Operation: One of the processing steps that materials undergo in a chemical process plant. Mixing and drying are examples of unit operations.

Feasibility Study: An analysis of a project to see if it can be carried out successfully. This is a common preliminary step in the planning of a new plant.

Research and Development (R&D): The gathering of the basic information needed for the design of a plant. Some of the information may be found in the library or learned from experts; the rest must be discovered in the laboratory.

Process Design: Making the decision on equipment to be used and developing all the other information needed for building a chemical process plant.

Flowsheet: A diagram that shows the equipment used and the steps by which a raw material is changed into a finished product. All process design is based on the flowsheet.

Instrumentation: The devices used for measuring or controlling a property such as temperature or pressure. The individual devices are called instruments. In an automobile, for example, the instrumentation includes the speedometer, the gasoline-level indicator, and the water-temperature indicator.

Plant Operation Engineer: The engineer in charge of a process plant after it is built. He or she may be the plant manager or may report directly to the plant manager.

Preventive Maintenance: The job of maintaining equipment in working order before it breaks down. In a chemical process plant the breakdown of an important piece of equipment may force the entire plant to be shut down, which can be very expensive in terms of lost production.

Consultant: An expert in some field who sells specialized knowledge to persons who need it.

Vocabulary Practice

1. What are the *chemical process industries*? How is the term usually abbreviated?

2. What is a *raw material*? Give two examples.

3. What is a *unit operation*? Is mixing a unit operation?

4. What is a *feasibility study*? When is one likely to be made?

5. What is *research and development*? How is the term abbreviated?

6. What is *process design*? Which comes first, R&D or process design?

7. What is a *flowsheet*? Who would use one?

8. What is *instrumentation* used for?

9. What is a *plant operation engineer*?

10. Name some instruments you are familiar with.

11. What is *preventive maintenance?* Why is it particularly important in chemical process plants?

12. What does a *consultant* do?

Most chemical engineers work in the *chemical process industries.* These include the plants that manufacture such things as food products, plastics, paper, fertilizers, petroleum products (gasoline, kerosene, fuel oil), synthetic (manmade) fibers such as nylon, and the basic chemicals used by many industries, such as acids, alkalis, and dyes. These are all industries in which *raw materials* are separated or changed into useful products.

Almost all chemical engineers are college-trained in mathematics and physics, with particular emphasis on chemistry. However, the basis of chemical engineering is the study of *unit operations.*

Before the first World War (1914 to 1918), all chemical process plants were designed and operated by chemists. (Even today, in some countries such as Germany, this work is done by specially trained chemists.) However, shortly after the war, three American college professors—Walker, Lewis, and McAdams—published a book based on principles common to all chemical process plants. They noted that in all plants materials were mixed together, heated, cooled, moved from place to place, and wanted materials were separated from wastes. Each of these steps was termed a unit operation, and the student was taught both the engineering principles that underlie each one, and the procedures used to design or select equipment for each operation.

Every chemical process consists of a number of sequential unit operations. Much of this book consists of descriptions of unit operations, so they will not be further discussed at this time.

How does a chemical process plant come into being? It starts with an idea—an idea for a completely new product, for improvement of an existing product, or for a way of producing an existing product at a lower cost. Ideas for completely new products usually come from a company's research laboratories but improvements on existing products may occur to almost anyone.

An 18th century plant for the manufacture of alum. The mineral alunite (A) was heated (BB) and extracted by water in ponds (CC). The liquid containing the alum was evaporated (G) and sent to settling basins (MM) where the alum crystallized and was removed.

Once the executives of a company have become interested in the idea of building a new plant, their first step is usually to call for a *feasibility study*. Such a study involves estimating production costs for the product as well as its potential market. Since essential engineering information is usually lacking, these estimates may contain major uncertainties.

If it appears that the plant will make a reasonable profit, the next step is to develop the engineering data that will be needed in designing it. This is the job of the *research and development* engineer. The R&D engineers who work for a CPI company are generally chemical engineers, although in large companies some mechanical, electrical, and civil engineers may also be employed.

R&D engineers do part of their work in the library with books and articles. They often work with other specialists, most often chemists, who are expert in some aspect of the problem. And they may do or direct some laboratory work themselves. But there is a great deal of difference between making a product in a laboratory and making it in a chemical plant. For example, penicillin was developed by growing a mold on a nutrient solution in a flask. The first commercial production was the result of doing the same thing in thousands of flasks.

When chemical engineers were called in to work on the problem, they devised a method of growing the mold in thousand-gallon tanks. Large quantities of sterile air were bubbled through the tanks to provide the oxygen the mold needed for growth. (The air had to be sterile so that no bacteria would grow in the solution.) In the flasks, the mold grew only on the surface where it could get oxygen from the air. But in the tanks, there was sufficient air so that the mold could grow beneath the surface as well. Within a short time production was so high—and the price so low—that the drug was widely available.

Because commercial production can be different from the laboratory process, the R&D engineer will often build and operate a model of the proposed plant in order to find out what kinds of problems may develop and how to solve them. When the research and development work is completed, enough information is available so that the original cost estimate can be refined to a fairly exact figure. Again, the company management has to decide whether to go ahead with the plant or to cancel the project.

Photo Courtesy Du Pont
Research and development engineers often build a scale model of a proposed plant to help predict and solve future problems.

If the company decides to go ahead, the next step is *process design*. In this stage, the chemical engineer decides what kinds of equipment will be needed for each unit operation and calculates the size of each item. He or she must also select the material that each equipment item is to be made of—usually metal, plastic, or glass—and contact various equipment manufacturers about prices.

One of the tools with which the process design engineer organizes all this information is the *flowsheet*. This is a diagram that shows what happens from the time a raw material comes into the plant to the time it emerges as the desired product. The R&D engineer will probably have made a simple flowsheet to help him or her understand the process, but the one made by the process design engineer will be much more complete. It will show all the pieces of equipment in the plant and how they are connected. The flowsheet will indicate the temperatures, pressures, and flows at each step of the process, and other things as well. One of these other things is the *instrumentation* that will be needed for operating the plant. Most processes in a CPI plant take place inside the equipment and it is only by using instruments that the operators can tell what is happening. If something might lead to a dangerous condition, the in-

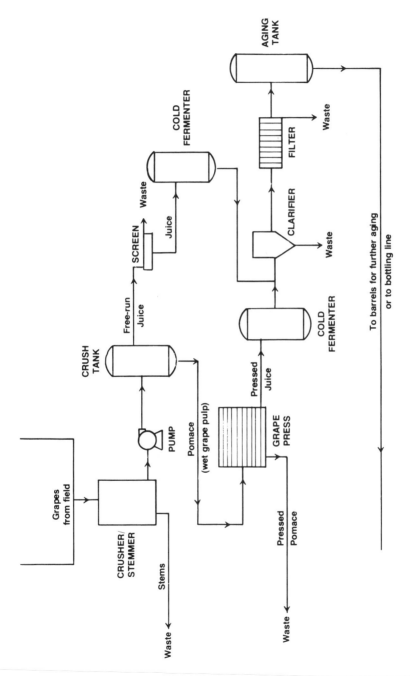

A flowsheet for a winemaking plant.

Photo Courtesy Du Pont
Instrumentation does so much of the work that it is possible to run a large chemical plant with only a few operators.

strumentation designer will generally provide flashing lights or ringing bells to call the operators' attention to the developing problem. If it cannot be solved immediately, the entire plant may have to be shut down.

In many plants, instruments not only indicate what is happening but also run the process automatically. A person walking through a modern chemical process plant for the first time is often surprised at how few people are working there. It is possible to run a very large plant with only a few operators and maintenance people because the instrumentation does so much of the work.

After the process design has been completed, the design engineer often supervises the building of the plant. Chemical process plants are usually built by specialized construction companies accustomed to working closely with process design engineers. When the plant is finished, chemical engineers are placed in charge of it to ensure its proper operation. They are known as *plant operation engineers* and are usually executives who spend most of their time at desks. Many people work under them: operators who run the plant on a day-to-day basis; maintenance personnel who keep the equipment operating; material handling personnel who move materials from place to place in the plant; cleaners, clerks, and others. Supervising equipment maintenance is an important part of the plant engineers' work. They must be sure that both spare parts and trained maintenance personnel are always available to prevent shutdown of the plant. The main priority is to anticipate and prevent machinery problems. Repairing equipment before it breaks down is called *preventive maintenance.* Sometimes instruments are installed to indicate if equipment is running hot or vibrating excessively. But more often, elaborate records are kept on critically important machines—records that show how long each part is expected to last so that new parts can be installed before the previous ones fail. Another responsibility of the plant operation engineer is to keep a supply of raw materials on hand, although some raw materials, like natural gas, are brought into the plant by pipeline and are always available. Another term for raw material is feedstock.

It is not uncommon for a CPI plant to run continuously for a year without a break. But some pieces of equipment, such as high-speed pumps, cannot be expected to run that long without being stopped for maintenance. In such cases, a spare piece of equipment will be installed permanently, with piping arranged so that either one can be used. Then, if the item must be repaired, the spare can be put into service to handle the load.

Although most chemical engineers work either in R&D, process design, or plant operation, some follow other careers. They may become college teachers to train new engineers or they may become salesmen of chemical-process equipment. An engineer who has become well known as an expert in some phase of chemical engineering may work as a *consultant*, charging high fees for solving problems too difficult for the average engineer. (College professors often earn extra money by working as consultants.) And recently, more and more engineers are taking jobs with government. These usually involve enforcing laws that relate to health or safety.

Discussion

1. Why is a plant that makes dynamite one of the chemical process industries?

2. Petroleum is the raw material for a large industry. List some products made from petroleum.

3. What stages might the chemical engineers of a chemical process company go through if they wanted to build a plant to make a new kind of fertilizer?

4. Most homes contain various kinds of equipment, particularly in the kitchen. How might an engineer use preventive maintenance to keep such equipment in working order?

5. What are some of the things done by the engineer in charge of plant operation?

6. Why is a consultant likely to be highly paid?

7. Most chemical engineers work in R&D, process design, and plant operation. What are some of the other jobs chemical engineers may do?

8. In your country, is the government passing laws for improved health and safety? Why might chemical engineers be hired to help enforce such laws?

Review

A. Look around the room and make a list of some things that are products of the CPI. (Hint: Don't forget your clothing and the objects in your pockets.)

B. If you go into a plant that makes automobiles you will find it full of workers. Why does a chemical process plant seem so empty of people when compared to an automobile plant?

C. Match the terms in the left column with the proper phrase in the right column. Only one definition is appropriate for each term.

1. CPI   _____ A step in a chemical process

2. Feasibility study   _____ Devices for measuring and controlling

3. Unit operation   _____ Chemical process industries

4. R&D   _____ Keeping equipment from breaking down

5. Flowsheet   _____ Collecting basic information for a new process

6. Instrumentation   _____ Diagram of a chemical process

7. Preventive maintenance   _____ Material that is made into a different product

8. Consultant   _____ Determining if a project can be successful

9. Raw material   _____ An expert in a special field

UNIT TWO
RESEARCH AND DEVELOPMENT

Special Terms

Laboratory: A place especially equipped for experimentation, testing, and/or analysis in a particular field of science or technology.

Flowmeter: An instrument used for measuring the flow of fluids (liquids or gases). Many different types are available.

Hopper: A container, usually funnel-shaped, for storing and delivering powdered or granular material. It is filled from the top, and the bottom is often equipped with a device for delivering measured quantities of the material.

Steam Jacket: A shell fashioned around a tank or other vessel. Steam is introduced into the space between the vessel and the shell, thereby heating the vessel and its contents.

Batch Process: A way of manufacturing chemical products. Measured quantities of materials are carried through a series of operations, step-by-step, to produce the final product.

Continuous Process: A way of manufacturing chemical products in large quantities. Raw materials are fed continuously into one end of the processing plant, flow through various operations, and emerge as the desired product. Continuous processes may run for months or years without stopping.

Proportioning Pump: A device usually consisting of several interconnected pumps. They are designed so that their outputs are adjustable, thereby permitting the ratios of materials discharged to be changed.

Heat Exchanger: A device for heating or cooling fluids. Steam is usually used for heating and cold water for cooling.
Pilot Plant: A miniature plant used for experimentation.
Shift Work: A way of staffing a plant or laboratory continuously for long periods of time. The workers are divided into groups; each group works at a different time.
Group Leader: The person in charge of a group of people. In experimental work, the leader is usually an engineer or scientist.
Progress Report: A description of the work of a group of researchers. Generally, a progress report is written each month.

Vocabulary Practice

1. What is done in a *laboratory?*

2. For what purpose is a *flowmeter* used?

3. What is a *hopper?* What kinds of devices are often used at the bottoms of hoppers?

4. Describe a *steam jacket.*

5. What is a *batch process?*

6. What is a *continuous process?*

7. Describe a *proportioning pump.*

8. What does a *heat exchanger* do? What is usually used in it for heating? For cooling?

9. What is a *pilot plant?*

10. Describe *shift work.*

11. What is a *group leader?*

12. What is a *progress report?* How is it usually presented?

Research and Development

The chemical process industries spend more money on research and development than do most other industries. As a result, we now use many kinds of products unheard of a few years ago. Countless items in our daily lives are different from those our parents used, because of this innovation. Much of our clothing is now made of synthetic fibers instead of natural materials such as wool or cotton. The toys our children play with are often made of plastics that replace wood or metal. And many of us drink instant coffee rather than brewing the beverage from ground coffee beans.

These kinds of products have come about through research and development in research *laboratories*. These laboratories are usually staffed by chemists who do their experimentation in the usual laboratory glassware. For example, when two materials must be mixed together, the chemist may do it with a glass rod or by merely shaking the container. The mixture can be heated by placing the container over a small gas burner or cooled by setting it in cold water. But many of the things that seem so easy in the laboratory are much harder to do in the plant. Even making the same product in the same way, but on a larger scale, presents many problems.

Let us look at a very simple process as the chemist does it and as it might be done in a chemical process plant. He or she (many chemists are women) takes a bottle of Chemical A from a shelf and pours the required quantity into a glass measure. The chemical is dumped into a flask and a second liquid, Chemical B, is measured and added in the same way. Chemical C, a powder, is weighed on a small laboratory scale and added to the two liquids. The chemist mixes the chemicals together by shaking the flask and heats the mixture over a small gas flame, with constant shaking. Finally, the mixture is rapidly cooled by placing the flask into a container of crushed ice. The chemist may have made a total quantity of a half-liter or less of product.

Now consider the same process carried out in a plant in batches of a thousand gallons. (Most chemical plants in English-speaking countries still use the old units such as pounds, feet, and gallons.) Instead of a glass flask, the container will be a thousand-gallon metal tank. Chemical A will not be in a bottle on a shelf—it will be in a storage tank. The proper amount of Chemical A will be added by pumping it from the storage tank through a *flowmeter* in-

Photo Courtesy Du Pont
Many of the products we use every day began in a research laboratory like this one.

to the processing tank. Flowmeters usually show flow in gallons per minute, so five-hundred gallons might be added at fifty gal/min for ten minutes. Chemical B will be pumped from its storage tank in the same way. (The operator will probably pump both liquids into the processing tank at the same time unless there is a danger in mixing them this way.) Adding Chemical C—the powder—is not as easy. If only a small amount is needed, it might be weighed into a container and dumped by hand into a mixture. If larger quantities are required, Chemical C will probably be stored in a *hopper* over the processing tank; the hopper will have some sort of measuring (usually weighing) device to ensure adding the proper amount.

Mixing a thousand-gallon tank cannot be done with a glass rod so the engineers will have provided a mechanical mixer, something like a ship's propeller, driven by an electric motor. The tank will probably be heated by steam supplied to a *steam jacket* surrounding the tank. The mixture in the tank is stirred throughout the heating period to make sure it is heated uniformly. To cool the mixture, the steam will be shut off and cold water pumped into the jacket. Water cannot be cooled below 32°F (0°C) without turning to ice, but colder temperatures can be achieved by using a solution of salt water called brine. Finally, the mixture will have to be pumped into another tank for storage or the next stage of the process.

What has just been described is an example of a *batch process* in which a given amount of chemicals is processed in some way to yield a quantity of product. Batch processes are commonly used when relatively small quantities of materials are handled and are particularly common when the materials are expensive. For example, batch processes are very often employed in the manufacture of drugs, dyes, and foods.

Another widely used way of handling materials is the *continuous process* in which materials are constantly fed into one end of the equipment and finished product comes continuously out of the other end. All petroleum refining, for example, is done by continuous processes. In a refinery, crude petroleum is pumped into one end of the plant and a continuous stream of gasoline, kerosene, and fuel oil pours out of the other end. Let us look at the simple process we have been describing (mixing quantities of Chemicals A, B, and C, heating, and cooling) and see how it might be done as a continuous process.

The two liquid chemicals are pumped by a *proportioning pump* into a small mixing tank, where Chemical C is continuously added by a solids feeding device. The mixed chemicals are continuously drawn off the bottom of the mixer and passed through a *heat exchanger* where they are steam heated. They then pass into another heat exchanger where they are chilled by cold water or brine. The products flow out of the end of the cold heat exchanger.

Photo Courtesy National Coal Association
This large scale pilot plant tests a process for making substitute natural gas from coal.

Because chemical plant equipment is so different from that used in the laboratory, one of the major jobs of R&D engineers is to decide what kinds of equipment must be used to carry out a commercial chemical process. They also determine the sizes of equipment needed—how big must the pumps be, how much power must the mixer have? Before designing the full-sized plant, the R&D engineer usually constructs a *pilot plant*—actually a small model of the final plant, containing small versions of the equipment. Pilot plants are particularly useful when designing continuous process plants which are so different from the research laboratory. (It is usually impossible to run a continuous process in the standard laboratory glassware available to the chemist.)

A continuous process pilot plant will usually run twenty-four hours a day with three or four groups of operators and engineers, each group working for eight hours. This is called *shift work*, and each group is called a shift. Most often, shifts work from 8 a.m. to 4 p.m., 4 p.m. to midnight, and midnight to 8 a.m. A fourth shift is needed if the plant is to run during weekends, although many pilot plants shut down at that time. Usually there is a *group leader* in charge of each shift. The group leader may be a chemical engineer, a chemist, or a specially trained operator. This arrangement makes pilot plant experimentation unattractive to many chemical engineers who prefer to work during the day and leave the evening and night shifts to specially trained operators. However, a pilot plant is often so complicated that engineers are required on all shifts.

Since the basic purpose of the pilot plant is to gather information, there are frequent changes of flowrates, pressures, and temperatures. R&D engineers are always looking for that combination of conditions that will enable them to produce the maximum amount of product at the minimum price. As information is gathered, it is passed along to the company's management. This may be done by memoranda and telephone calls but in most companies, once a month, the R&D engineers write all they have learned during the past month in a *progress report*. These become their main record of accomplishment. The purpose of R&D is to gather information; since a company's management judges R&D engineers by the reports they submit, a great deal of work goes into the reports' preparation. When the research and development project is completed, information in the various progress reports is consolidated into a final report that details everything learned during the research. This final report is invaluable to the process design engineers who will design the full-scale plant.

There is one thing about R&D that many engineers find frustrating: a project is seldom finished. As with all research, there are always more ideas than time or manpower. Eventually, the work must end, even if the best possible design has not been reached. Otherwise no process would ever get into full-scale production. The decision to end a project is usually made by the head of the research laboratories in consultation with the executives of the company.

Discussion

1. What kind of work is done in a laboratory?

2. What kind of specialists usually do laboratory work?

3. Of what material is most laboratory equipment made?

4. How might a laboratory chemist mix materials?

5. What could a chemist use for heating or cooling chemicals in the laboratory?

6. What kinds of equipment might a chemical engineer design for mixing large quantities of materials? What might he use to measure these materials?

7. What is the device engineers use to heat tanks of materials?

8. What is a solution of salt in water called?

9. How does a batch process differ from a continuous process?

10. What kind of equipment is used for the continuous heating or cooling of streams of chemical materials?

11. What kind of equipment might a chemical engineer use for experimenting on a continuous process? Why doesn't a chemist usually run continuous process in the laboratory?

12. Why is shift work necessary in running a pilot plant?

13. What are the usual time periods for each shift?

14. Why is pilot plant work unattractive to some engineers?

15. What do chemical engineers look for when running a pilot plant?

16. How do R&D engineers pass on information to a company's management?

17. Who usually decides when a research project should be ended?

Review

A. List as many kinds of things as you can think of that are now made of plastics but were formerly made of natural materials.

B. Employees in the chemical process industries often have to work in shifts. What are some other kinds of jobs that require shift work?

C. Complete the following sentences with an appropriate word or phrase.

1. Ideas for new products are generally developed in a
   _____.

2. A person experimenting in a laboratory is likely to be a
   _____.

3. A chemical engineer uses a _____ to measure large amounts of liquids.

4. A _____ is used for storing and delivering powdered or granular materials.

5. A process plant mixer may look something like a ship's
   _____.

6. A _____ is often used to heat materials in a tank.

7. Batch processes are often used to process materials that are
   _____.

8. Quantities of materials may be added in fixed ratios to a process by using a _____.

9. A _____ is employed for continuous heating or cooling of materials.

10. A _____ is a miniature plant used for experimentation.

11. A _____ plant will generally operate twenty-four hours a day.

12. People who operate continuous process plants generally do _____ work.

13. _____ shifts are usually needed if a plant is to work on weekends.

14. The basic purpose of a pilot plant is to _____.

15. Engineers write _____ to inform management of their findings.

16. Progress reports are important because_____
_____.

UNIT THREE
PROCESS DESIGN

Special Terms

Scaleup: A mathematical technique whereby data from experiments on small-scale equipment can be used in designing plant equipment.

Vessels: The hollow structures containing liquid or gas. Tanks are the most common vessels used in CPI plants.

Kettle: A tank, usually open at the top, fitted for heating and cooling its contents.

Tank Farm: An area of a chemical process plant where there are a number of large tanks used for storing raw material or finished product.

Pipe Flange: A heavy metal disk which attaches to the end of a pipe to connect one pipe to another.

Gasket: An elastic material used for making joints tight. It is usually used between two mating pipe flanges.

Pipe Fitting: A specially made short length of pipe for such purposes as changing direction, joining pipe at various angles, or connecting pipe of unequal sizes.

Pipe Elbow (Ell): A pipe fitting used to change pipe direction by 90 degrees or by 45 degrees.

Pipe Tee: A T-shaped fitting used to connect pipes.

Pipe Cross: A pipe fitting used to connect four lengths of pipe at right angles to each other.

Pipe Reducing Fitting (Reducer): A pipe fitting used to join two pipes of unequal diameters. It comes in two types—eccentric and concentric.

Pipe Lateral: A fitting used to join two pipes at an acute angle.

Pressure Vessel: A closed vessel especially designed to contain its contents under pressure. Generally, any internal pressure above one atmosphere (100 kilopascals) requires a pressure vessel.

Pressure Vessel Code: A set of rules for designing and testing pressure vessels. In the United States the American Society of Mechanical Engineers (ASME) has devised such a set of rules—the ASME Pressure Vessel Code. Many governments require that pressure vessels conform to this code, thus giving it the force of law.

Materials of Construction: A term comprising various materials, such as metals and plastics, used in building structures and equipment. Selection of the proper material of construction for equipment is one of the jobs of the design engineer.

Corrosion: The attack on metals by various chemicals. Materials, such as acids, that attack metals are said to be corrosive. Special materials that withstand such attack are called corrosion resistant.

Alloy: A combination of metals (and sometimes other elements) that are melted together to produce another metal with special properties such as high strength or corrosion resistance.

Exotic Metals: The relatively rare and expensive metals that have special properties. They include titanium, tantalum, and zirconium.

Capital Investment: The total cost of land, building, and equipment for a plant or facility.

Operating Cost: The expense involved in running a plant. It includes such things as raw material, labor, maintenance, and replacement.

Return on Investment (ROI): A relationship between the cost of a plant and the profit made from that plant.

Vocabulary Practice

1. What is meant by *scaleup*?

2. What is a *vessel*?

3. What is a *kettle*?

4. Describe a *tank farm*.

5. What is a *pipe flange*? What is it used for?

6. What is a *gasket*?

7. What is a *pipe fitting*?

8. Describe a *pipe elbow*.

9. What is a *pipe tee* used for?

10. How is a *pipe cross* used?

11. What are *pipe reducing fittings*? What are they usually called? What are the two types of *reducers*?

12. What is a *pipe lateral*?

13. What is a *pressure vessel*? What pressures require such vessels?

14. Explain the meaning of *pressure vessel code*. Name such a code.

15. What are *materials of construction*? Who usually selects them?

16. What is *corrosion*? What are materials that resist corrosion called?

17. What is an *alloy*?

18. What are *exotic metals*? Name some.

19. What is meant by *capital investment*?

20. What is an *operating cost*? What items are included in this term?

21. What is *return on investment*? How is it abbreviated?

Photo Courtesy Dow Chemical
A Dow Chemical crude oil processing plant under construction.

Process Design

Many chemical engineers feel that process design is the most interesting kind of work in their profession. Certainly the training chemical engineers receive in college is more concerned with design than with any other aspect of chemical engineering. Most of the information needed by the process design engineer is generated during the R&D phase, particularly from pilot plant data. But that information is based on small-sized equipment, whereas the production plant will contain full-sized equipment. Adapting that data is known as *scaleup*. This is always a mathematical technique but the exact procedures used depend on the type of equipment being scaled-up. Procedures for doing the job are available in textbooks and handbooks and may frequently be found in manufacturers' catalogues and other such literature.

Almost any chemical process plant includes a vast array of pipes and *vessels*, with another odd-shaped piece of equipment here and there. Open vessels are usually called tanks though, when heated, they are sometimes called *kettles*. Plants that process large quantities of liquid raw materials (such as petroleum refining plants) often have a number of enormous tanks arranged in an area known as a *tank farm*. When the contents of these tanks are flam-

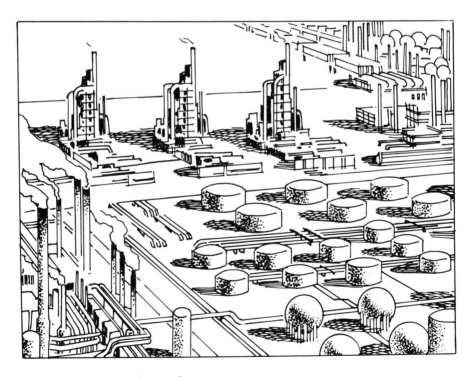

A tank farm at a petroleum refinery.

mable, there are usually embankments of earth called dikes around them; if the tanks leak and the material catches fire, the flaming liquid cannot spread over the entire plant (as it did in early disasters).

Piping is an important part of any chemical process plant because there is so much of it. Most home piping is of small size and is put together by screw threads cut into the pipe. Screwing sections of pipe together is easy to do with small-sized pipe because it can be easily handled. But the pipe used in process plants generally ranges in size from two to twelve inches in diameter and a section of such pipe may weigh many hundreds of pounds or kilograms.

Large-diameter pipe is put together by means of welding or *pipe flanges*. When flanges are used they are usually welded to the ends of the pipe. The flanges have a ring of holes around them so that two flanges can be bolted together with a *gasket* in between. Flanges have two advantages: first, welding can be done on the ground or in a shop; second, the pipe can be easily disassembled for inspection or replacement. The disadvantages are that the joints sometimes leak and the flanges are expensive. Welded pipe is harder to assemble because the welds must often be made while the pipe is high in the air or in awkward positions. A good welded joint will not leak and welding is a cheaper way of assembling pipe than flanges. However, the pipe must be cut and rewelded if it has to be opened for inspection or if it is necessary to replace faulty sections.

Pipe is made with walls of various thicknesses—thicker walls are required to carry material under higher pressure. Part of the design engineer's job is to calculate the pressures a pipe will have to withstand and to specify walls of the proper thickness.

There are a number of *pipe fittings* used with all piping systems. When a pipe must change 90 degrees in direction, a *pipe elbow*, or 90-degree *ell*, is employed; 45-degree ells change the direction by 45 degrees. When one pipe joins another at right angles, the juncture is made with a *pipe tee*. A *pipe cross* joins four pieces of pipe. Pipe of two different sizes can be connected by *reducers*, or *pipe reducing fittings*. When pipes must be joined at a sharp angle, a *pipe lateral* fitting is used.

As with pipes, vessels that withstand high pressure must have extra heavy walls. These are known as *pressure vessels* and must be carefully designed and constructed. Failure in pressure vessels may result in explosion with large pieces of the vessel flying around the

TEE, STRAIGHT

90° ELBOW, STRAIGHT

CROSS, STRAIGHT

45° ELBOW

ECCENTRIC REDUCER

Flanged Pipe Fittings

plant—a circumstance exceedingly dangerous to personnel and destructive to equipment. Therefore the design and construction of pressure vessels (usually defined as those containing above 15 lb/sq. in., or 1 kilogram force/sq. cm., or 100 kilopascals) is usually regulated by laws known as *pressure vessel codes.* Any governmental body may write its own code. In the United States most areas adopt the pressure vessel code of the American Society of Mechanical Engineers (ASME) which is then enforced by Governmental pressure vessel inspectors. This code defines the way in which a design engineer must calculate vessel wall thicknesses and shapes, and also sets up fabrication procedures to be followed during construction. Welding, for example, must be done with extreme care and skill. Generally, welders who work on pressure vessels must pass a special test and are then known as certified welders. After a vessel has been designed and built, it must pass a pressure test conducted under conditions spelled out in the pressure vessel code. Because of the complications surrounding pressure vessels, the process design engineer usually designs only large or specialized vessels. Smaller ones can be purchased in standard sizes from manufacturers who specialize in such equipment. These vessels carry a special nameplate certifying that they were built according to the code and that they have passed the required tests.

Designing process plant equipment requires more than scaleup or following design codes; it also means specifying the materials to be used for equipment. Actually, the material must be selected before the equipment is designed because the design is related to the strength of the various *materials of construction.* Most chemical process equipment is built from ordinary steel, called mild steel or carbon steel. But many chemicals attack mild steel, causing *corrosion.* When this is a problem, the design engineer usually turns to stainless steel, an *alloy* of iron, nickel, and chromium. Some stainless steels contain other metals, such as molybdenum, that give the alloys greater strength, resistance to corrosion, or another desirable property. When corrosion problems are too severe for stainless steel, the engineer may choose alloys that contain large quantities of nickel known as high-nickel alloys. These special alloys are usually sold under trade names (some are Hastelloy, Incoloy, and Inconel) and may cost as much as several dollars a pound. Some process conditions require more specialized metals like titanium, tantalum, or zirconium. These are called *exotic metals* and are also very expensive.

In addition to metals, the design engineer has a large number of nonmetals to choose from. These include glass, plastics, graphite, and ceramics—alone or in combinations like glass-lined steel or rubber-lined plastic. The graphite used in process plants is usually impregnated with plastic and is known as impervious graphite. There are limitations to the use of nonmetals; most are not as strong as metals so they cannot be used for high pressure equipment, nor can they withstand high temperatures.

A process plant is usually designed in several stages. After the research and development has been done, and before the decision is made to build a full-sized plant, a final economic evaluation must be made. Of course, economic evaluations were made before and during the R&D work but they could be little more than educated guesses. The first time there is enough information to make a firm estimate of the cost of the plant and the cost of producing the product is when R&D is complete.

The cost of the final product is based on two things: *capital investment* and *operating cost.* The first is the cost of the land, the plant, and its auxiliaries such as roads, railroad spurs, electrical power lines, and sewage treatment facilities. Included here is the cost of equipment, materials, supervision, labor, and design. The total of these costs—a large sum of money that must be available at one time—is the capital investment. The second, operating cost, includes the expense of running the plant after it is built—raw materials, labor and supervision, maintenance, shipping, and so on.

To determine whether or not an operation will be profitable, the design engineer must calculate the rate of return or *return on investment* (ROI) as follows:

$$\% \, \text{ROI} = \frac{\text{Net annual income - net annual depreciation}}{\text{Capital investment + R \& D costs}} \times 100$$

Annual depreciation is the annual loss in value of the plant and its equipment owing to wear and obsolescence. In the chemical process industries, it is common to require that a project have an estimated ROI of at least 20% after income tax before it will be approved for construction.

Discussion

1. What aspect of the work do most chemical engineering college courses emphasize?

2. When does the design engineer get most of the information he or she needs for design?

3. What does a design engineer do to translate the pilot plant information into data for the full-scale plant?

4. Describe something of the appearance of a chemical process plant.

5. What does a tank farm look like?

6. How are pipes used in the home usually connected?

7. How are pipes connected in a chemical process plant? Why are these techniques used?

8. What are the advantages of connecting pipe by using flanges? By welding? What are the disadvantages of each method?

9. What are pipe fittings used for?

10. Name and describe five kinds of pipe fittings.

11. Why are pressure vessels used? What kind of pressure do they have to withstand?

12. Why may pressure vessels be dangerous if they are not properly designed and built?

13. Who writes pressure vessel codes? What is the most important pressure vessel code used in the United States?

14. What is special about the welders who work on pressure vessels?

15. Must the process design engineer design all the pressure vessels to be used in a plant? Why?

16. How can you tell if a purchased pressure vessel is safe?

17. What metals are used as materials of construction in chemical process plants?

18. What goes into the alloy known as stainless steel?

19. Name some high-nickel alloys.

20. What materials are called exotic metals?

21. Name several materials of construction that are not metals.

22. What is done to graphite to make impervious graphite?

23. What are some limitations of the nonmetallic materials of construction?

24. At what stages during the time a new process is being developed are economic evaluations usually made? When can such an evaluation best be done?

25. What are the factors that go into the cost of the product of a CPI plant?

26. How does an engineer calculate return on investment?

27. What is a commonly required ROI before a plant will be approved for construction?

Review

A. Some commonly used pipe fittings are shown above. Name each one and explain how it might be used in a chemical process plant.

B. Corrosion is an important problem in chemical plant design. From your own experience, discuss the kinds of chemicals likely to cause corrosion. What kinds of corrosion might occur at home? In an automobile?

C. The ASME Pressure Vessel Code is the one most commonly used in the United States. What codes are used in your country? Discuss the reasons for having such codes.

UNIT FOUR
PLANT OPERATION

Special Terms

Foreman: A specially trained worker selected, because of work skills and leadership abilities, to supervise other workers.

Plant Efficiency: A comparison between the amount of product made in a plant and the cost of making it. The more efficient a plant is, the lower the cost of making a given quantity of product.

Plant Throughput: The amount of product manufactured in a particular length of time. A plant's throughput, or output, is usually given as tons/day or tons/yr.

Startup: The process by which a plant begins operations. The initial startup—the first time operation is attempted—may take months before all the equipment and the control system are working satisfactorily.

Shutdown: The process by which plant operation is stopped. Pieces of equipment must be stopped in a particular order, and sometimes in special ways, to avoid problems.

Column: A tall, narrow, closed vessel used for a particular purpose. It may have a height eight or ten times its diameter, and may be 100 feet (30 meters) tall, making it a striking feature of a process plant.

Reactor: A vessel (usually closed) in which a chemical reaction takes place. Reactors which work under high pressures—pressure vessels—are usually equipped for heating or cooling their contents.

Purging: The process of clearing liquids or gases out of pipelines and equipment. Flammable or dangerous materials are usually purged by displacing them with a safe material.

Instrument Tuning: The process of adjusting instruments that automatically control a plant's operation.

Lost-time Accident: A mishap that requires a worker to be away from the job for a day or more. Records of plant safety are usually based on the number of lost-time accidents.

Quality Control: The process of assuring that a plant's products have the required physical and chemical characteristics. The qualities checked may include such things as purity, strength, and color.

Rotating Shifts: A form of shift work in which each group of workers is changed at regular intervals from day work to evening work to night work.

Insulation: A material that provides a barrier to the passage of heat. It is used to keep hot things hot and cold things cold. Insulation contributes to plant safety by preventing contact with dangerously hot surfaces.

Vocabulary Practice

1. What is a *foreman*? On what basis are they usually selected?

2. What is *plant efficiency*? How is it usually calculated?

3. How is *plant throughput* determined?

4. What is meant by *startup*? What may be special about the first one?

5. What is *shutdown*?

6. Describe a *column*.

7. What is a *reactor*? What is it equipped to do?

8. What is meant by *purging*? How is equipment purged?

9. What is meant by *instrument tuning*?

10. What is a *lost-time accident*? Why is it recorded?

11. What is *quality control*? What may it check?

12. What are *rotating shifts*?

13. What is *insulation*? How does it contribute to safety?

Plant Operation

The function of the chemical engineer involved in plant operation is to see that the plant manufactures product. This sounds simple but includes a great many details. The engineer's two big responsibilities are the plant itself and the people who run it. The operations engineer is in charge of plant maintenance which is a continuing job in any chemical process plant, large or small. He or she is also responsible for having trained personnel on hand for plant operation. Since plant production often varies with the country's economy and sales of the plant's products, the size of the work force must also vary. Personnel are hired and laid off at various times. Sometimes workers can be recalled but often they have taken other jobs and new people must be hired and trained. In addition, personnel are discharged, quit, or retire, and some are sick or injured. Many CPI plants are unionized so operations engineers must be skilled in dealing with unions, both in day-to-day matters and in bargaining when union contracts expire.

The plant operation engineer is also responsible for the selection and training of *foremen*. A foreman (who may be a man or woman) is a specially trained person who is in charge of a group of workers. The plant operations engineer is always on the lookout for skillful workers with leadership qualities. A foreman is part of the plant's management personnel and may advance to higher levels of management.

An ongoing occupation of plant operation engineers is that of increasing the *efficiency* and *throughput* of their plants. Efficiency of plant operation is generally defined as producing product of the

Photo Courtesy Dow Chemical
An engineer in a chemical plant takes a reading on a reactor.

required quality at the lowest possible cost. Sometimes, if there is a shortage of the product made by the plant, it may be important to increase the throughput even at increased cost. Hence, plant managers frequently experiment to find ways of increasing both efficiency and throughput. In theory, every process plant should always operate at the maximum efficiency possible, consistent with the required throughput.

Most large chemical process plants operate continuously; while the plant is running there is often little for plant personnel to do. Problems occur when some piece of equipment operates improperly or breaks down. But there are two occasions when plant personnel are always very busy: *startup* and *shutdown*. At the beginning of startup nothing is operating, so valves must be opened in the proper sequence to start the flow of materials; steam and cooling water must be started as required; *columns* and *reactors* must be brought to their proper operating temperatures. Some plants may take as much as a day before operating normally. Shutdown takes less time but also requires continuous attention from all plant personnel. This is particularly true of emergency shutdowns caused by equipment failures or by accidents such as fire and explosion. During shutdown it is often necessary to clear liquids or gases out of pipelines and equipment, particularly if they are flammable or corrosive. This is usually accomplished by displacing the dangerous material with a safer one—a process known as *purging*.

Perhaps the most laborious job in any plant is the very first startup. Inevitably, equipment is found to be defective or improperly installed. Pipe joints leak, pumps are improperly wired, and valves are jammed so they cannot be opened or closed. Often it takes several months of steady work before everything is corrected and the plant can run as it was designed to. Much of the work during an initial startup is devoted to what is known as *instrument tuning*. Most instruments used to make a plant run automatically have several adjustment knobs that must be set properly for the devices to work. These adjustments interact with each other—that is, changing one will change the action of the others. This means considerable trial and error before all are brought to the proper position.

Although chemical process plants use many dangerous materials, their accident rates are quite low as compared to other industries. Perhaps it is because the plants handle dangerous materials that everyone tends to be extremely conscious of safety. CPI plants generally carry on extensive safety campaigns. In many areas of a plant everyone, including management personnel and visitors, is required to wear a safety helmet (usually called a hard hat), or safety goggles, or both. Special safety showers and eyewash stations are provided to furnish emergency aid to any worker sprayed by a dangerous chemical. If poisonous gases are used, gas masks are kept

Photos Courtesy Dow Chemical

Safety is a primary concern of the chemical process industries. Above, laboratory animals used for testing the presence of air pollutants and below, a safety meeting held for Spanish-speaking workers at a Dow Chemical plant in Texas.

available. Every plant keeps bandages, burn dressings, and first-aid equipment readily available. Since many of the materials used in the CPI are highly flammable, there is always danger of fire. Smaller plants have fire-alarm boxes connected to the local fire department; large plants often have their own fire departments.

One way to determine the safety record of a plant over a period of time is to calculate the number of *lost-time accidents*—those serious enough for the one injured to be unable to work for a day or more. Many plants have large signs at the plant entrance indicating the number of days since the last lost-time accident.

Another plant responsibility is that of *quality control*. Every product of the CPI must meet certain specifications before it is sold. Enforcing these is the job of the quality control laboratory and its personnel. They check raw materials bought by the plant to determine if they are of the proper grade and check samples of product at specified times to see if the quality is being maintained. Quality-control personnel usually report only to the plant manager—not to other production supervisors—so that their standards and judgment cannot be influenced by production personnel.

Photo Courtesy Dow Chemical
The flexural strength of a cured resin is tested as part of quality control.

Since continuous process plants are generally staffed by four shifts, there must be four workers available for each production job. Consequently a plant's workforce is much larger than can be seen—only a quarter of the workers are in the plant at any one time. Most workers prefer the day shift, so the evening and night shifts are compensated by higher rates of pay. But shifts are usually changed every few weeks so that each worker has an equal share of the desirable and undesirable conditions; this is known as *rotating shifts*. Scheduling the four shifts can be very complicated and take up a considerable part of the operations engineer's time.

There are still some parts of the world where energy is comparatively cheap, but in most places it is a considerable part of the cost of producing chemical products. Energy conservation has become a major way of decreasing costs in most process plants. Hot pipes and equipment have always been insulated because they would otherwise be a safety hazard, but the modern trend is to use much more *insulation* in order to conserve heat. Chemical reactors frequently operate at high temperatures so they are prime candidates for thick insulation. High pressure steam lines also need heavy layers of insulation. In older plants it may be difficult to use as much insulation as is now desirable because the equipment and pipelines are so close together that there is insufficient room. There are some newer insulation materials that provide good protection with relatively thin layers, but they are expensive.

Discussion

1. What is the primary job of plant operation engineers?

2. What are some of their responsibilities?

3. What kinds of personnel problems does the plant engineer deal with?

4. What kind of worker might become a foreman?

5. Are foremen part of management?

6. When is it desirable to increase plant throughput even at lower efficiency?

7. Is a plant always operated at its highest efficiency?

8. What are the most frequent causes of operating problems in process plants?

9. When are plant personnel busiest?

10. What must be done during a plant startup?

11. What things must be done during a plant shutdown?

12. What are some types of process plant equipment brought up to operating temperature during a startup?

13. What may cause an emergency shutdown?

14. Why do pipelines and equipment sometimes have to be purged? How is this done?

15. What kinds of problems may occur during a plant's first startup?

16. Why is instrument tuning needed during an initial startup?

17. Do chemical process plants have high accident rates? What may be a reason for this?

18. What kinds of safety equipment do most plants have?

19. How do CPI plants fight fire?

20. Discuss lost-time accidents.

21. What is the purpose of quality control?

22. To whom do quality control personnel usually report? Why?

23. Usually, few people can be seen working in a process plant. Explain why the actual work force is much larger.

24. What is the purpose of having rotating shifts in a plant?

25. What are the two main reasons for using insulation?

26. Why is it difficult to use enough insulation in many older plants?

Review

A. The last three units have discussed the role of the chemical engineer in research and development, process design, and plant operations. Discuss the area of work that interests you the most and explain why.

B. From your own experience in chemical plants, discuss the ways in which engineers ensure safety.

C. Complete the following sentences with the proper word or phrase.

1. The initial _____ is the first time a plant is operated.

2. A _____ is a tall, narrow vessel.

3. Chemical reactions take place in a _____, usually under high pressure.

4. Stopping a plant requires orderly _____ procedures.

5. _____ is used to conserve heat energy.

6. A _____ supervises a group of workers in a plant.

7. The _____ laboratory is used to make sure that the plant's product meets specifications.

8. One reason for putting _____ on equipment is to protect workers against burns.

9. A _____ results in a day or more out of work.

10. _____ is the relationship between the amount of product made by a plant and the cost of making it.

11. If more product is needed, engineers must increase the plant's _____.

12. Adjusting control instruments to give optimum results is called _____.

13. If someone works during the day for a week, during the evening for another week, and at night for the third week, he or she is on a _____.

UNIT FIVE
CONTROLLING THE PLANT:
INSTRUMENTATION

Special Terms

Thermometer: An instrument used for measuring temperature. One type is based on the expansion of liquid in a graduated transparent tube.

Thermocouple: An electrical instrument that measures temperature. It is based on the electrical flow that occurs in an electrical circuit made up of two dissimilar metals when the two junctions between the metals are at different temperatures.

Calibrate: A means of standardizing, adjusting, or checking the graduations of a quantitative measuring instrument.

Thermocouple Well: A protective tube that isolates a thermocouple from the material whose temperature is being measured. It is frequently used to protect the thermocouple from corrosive liquids or gases. When a similar tube is used to protect a thermometer, it is called a *thermometer well.*

Pyrometer: A device that measures temperature by detecting the radiation from a hot object.

Bourdon-tube Pressure Gage: The type of pressure gage most commonly used. It is a curved metal tube, closed at one end and subjected to the pressure being measured at the other end. Internal pressure tends to straighten the tube and this movement is amplified and displayed on the graduated face of the gage.

Manometer: A simple pressure-measuring instrument in which pressure raises the level of liquid in a transparent tube.

Rotameter (Variable-area Flowmeter): A type of flowmeter in which fluid flows upward through a tapered tube and lifts a float or plummet to a point at which the force of the flow counterbalances the weight of the plummet.

Sight Glass (Sight-flow Glass): A transparent tube inserted into a pipeline to enable an observer to see the flow in a pipe.

Indicator (Indicating Instrument): An instrument that shows the measure of a particular variable such as temperature, pressure, or flow.

Recorder (Recording Instrument): An instrument that makes a record of the measure of a particular variable.

Controller (Control Instrument): An instrument that measures a variable and uses this measurement to control a process.

Control Valve: A valve that can be opened or closed (or set anywhere in between) by a control instrument. It is the combination of control instrument and control valve that accounts for most of the automatic operation in process plants.

Control Room: A room in which many of the instruments of a chemical process plant are located. This centralization of instruments enables one or two operators to run the entire plant.

Vocabulary Practice

1. What is a *thermometer*?

2. What is a *thermocouple*? How does it work?

3. What does *calibrate* mean?

4. What is a *thermocouple well*? Why might one be used in a process plant?

5. What is a *pyrometer*?

6. For what is a *bourdon-tube pressure gage* used? How does it work?

7. What does a *manometer* do?

8. What is a *rotameter*? How does it work?

9. What is meant by *sight glass*? What is another name for a sight glass?

10. What is an *indicating instrument*? By what other name is it known?

11. What is a *recording instrument*? What else is it called?

12. What is a *control instrument*?

13. For what purpose is a *control valve* used?

14. What is a *control room*?

Controlling the Plant: Instrumentation

We have seen how important a plant's instrumentation is in enabling it to function properly. Now let us take a closer look at the actual equipment that is used for this purpose. In any process plant the three basic variables to be controlled, if the plant is to operate properly, are temperature, pressure, and flow.

• **Temperature:** The two basic instruments for measuring temperature are the *thermometer* and the *thermocouple*. The basic design of the thermometer is familiar. It consists of a tube, usually glass, which contains a liquid that expands or contracts depending on the temperature to which it is exposed. As it expands it rises in the tube—the height to which it rises is *calibrated* to indicate temperature. Depending on the temperatures to be measured, the liquid in the tube is usually colored alcohol or mercury.

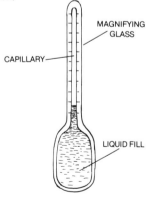

Thermometer

The second most common temperature-measuring device is the thermocouple. It is based on the discovery in 1821, by the German physicist, Thomas J. Seebeck, that an electric current flows in a continuous circuit composed of two different metals if the two junctions between the two metals are at different temperatures. One junction is placed at the point where the temperature is to be measured; the other is usually in the instrument used to measure the current. The current that flows in the circuit depends on the difference between the two temperatures. Since the temperature in the instrument is easily found (by using a thermometer, for example), the other temperature can be calculated. Some advanced thermocouple instruments are equipped to do the calculation automatically—on these, the desired temperature can be read directly. Since the output of the thermocouple is electrical, it is frequently chosen to be used with other electrical or electronic instruments.

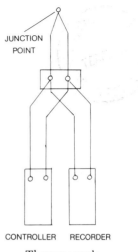

Thermocouple

If a thermocouple is to be placed in a pipe or vessel, it is often necessary to protect it from the corrosive or abrasive material being processed. In such cases, it is placed in a protective tube known as a *thermocouple well*. Similar wells are used with thermometers and are known as *thermometer wells.*

Thermometers and thermocouples are not suited to the very high temperatures inside furnaces—the heat would destroy them. These temperatures are usually measured with instruments called *pyrometers*. Hot objects radiate energy and the pyrometer measures this energy. The energy radiated by a hot object varies approximately with the fourth power of the absolute temperature of the object, and the pyrometer is calibrated on this principle. There are some other less common devices used to measure temperature. For example, the electrical resistance of a wire varies with its temperature so it is possible to use this effect to measure temperature. Devices based on the variation of resistance are called resistance thermometers. Other types of temperature-measuring devices are based on the expansion and contraction of solids under varying temperatures.

• **Pressure:** There are literally dozens of different types of instruments for measuring pressure. The most common is the *bourdon-tube pressure gage*. This consists of a curved metal tube that is closed at one end and open at the other. The open end is exposed to the pressure to be measured. As the pressure in the closed tube increases, it tends to straighten the tube. Since the metal of the tube is elastic, the closed end moves as the pressure within the tube changes. This slight movement is amplified by a system of levers and gears, and is used to move a pointer over a graduated scale.

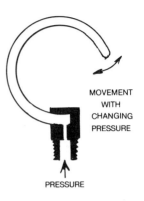

MOVEMENT
WITH
CHANGING
PRESSURE

PRESSURE

Bourdon Tube

Instead of the bourdon tube, a pressure gage may use a flexible plate, and is called a diaphragm gage. Or a corrugated tube called a bellows may be used to make what is

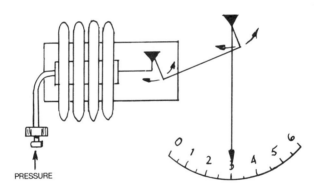

PRESSURE

Diaphragm Gage

called a bellows gage. Here again, movement of the diaphragm or bellows due to a change of pressure is translated into movement of a pointer on a dial. There are numerous other types of pressure-measuring instruments. For example, those used for high vacuum may measure the thermal conductivity of the small amount of gas remaining in the space whose pressure is to be measured. Those based on the electrical characteristics of materials in a strained or un-

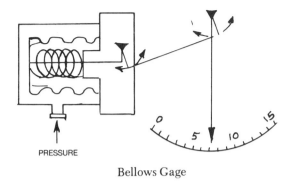

Bellows Gage

strained condition are called strain gages and are becoming increasingly common with electronic instrumentation.

Another common variety of instrument used for pressure measurement is the *manometer*. This uses a column of liquid (usually water or mercury) which is exposed to pressure at one end. The height to which the liquid rises in the tube is a measure of the pressure. Manometers are used mainly to measure small pressure changes. If the pressure is very low it may be measured by a manometer called a draft gage. Pressures measured by manometers may be shown in the conventional units of

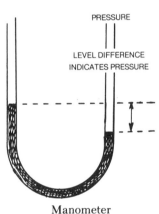

Manometer

pounds per square inch, or in pascals, or they may be simply given as the height of the column of liquid—inches or centimeters of water or mercury.

• **Flow:** The most common means of measuring rate of flow of a fluid is based on the principle that if there is a contraction or constriction (a section of reduced diameter) in a pipe, the pressure of a fluid flowing downstream from the contraction will be lower than that upstream from the contraction. This pressure differential is proportional to the rate of flow. The simplest form of such a flowmeter is known as an orifice-plate meter. It consists simply of a plate with a hole in it, inserted into the pipeline, usually between

two flanges. The hole in the plate is called an
orifice and is smaller than the internal diameter
of the pipe. Small openings, called taps, are
made in the pipe on both sides of the orifice
plate, and these are connected to some kind of
pressure-measuring device—most often a manometer. The manometer reads the differential
in pressure on both sides of the orifice and may
be calibrated to read in terms of rate of flow.
Orifice plates cannot be used if the liquid being
measured contains solids because they will be
trapped by the plate. The trapped solids inter-
fere with the calibration of the flowmeter and
may even clog the pipe. An orifice plate also
causes pressure losses and a consequent increase
in pumping costs.

Orifice Plates

Instead of an abrupt contraction in the pipe size, as is caused
by an orifice plate, the contraction may be gradual. Special fittings
(called venturi tubes) are used, by which the pipe is gradually nar-
rowed and gradually enlarged. The pressure taps are placed at the
largest part of the venturi tube (before it narrows), and at the
smallest diameter. These tubes overcome most of the problems of
the orifice-plate meter, but the venturi tubes themselves are rather
expensive, particularly when they must be made from corrosion-resistant materials such as
stainless steel. Although
there is a decrease in pressure at the narrowest part
of the venturi tube, the
pressure is almost entirely
recovered as the liquid
passes through the gradual
enlargement, so that added
pumping costs are mini-
mal. Another type of flowmeter is the *rotameter*, or
variable-area meter. The
liquid to be measured is
passed upward through a
tapered tube with the small-

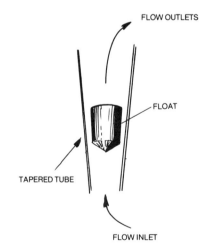

Rotameter (Variable-area Flowmeter)

est part of the taper at the bottom. In the tube is a small metal plummet, or float. (Although usually called a float, it is made of metal and is denser than the fluid to be measured.) As liquid flows up the tube, it lifts the float to a position where the weight of the float (a downward force) is exactly balanced by the lifting effect of the liquid in the space between the edge of the float and the inside of the tapered tube.

The ingenuity of engineers is great, and many other types of flowmeters have been invented and are in use. The types described above are simply the most common.

Sometimes it is important to know whether or not a liquid is flowing in a pipe, without needing to know its velocity. For this purpose, a *sight glass* (or *sight-flow glass*) is employed. This is simply a glass tube mounted between metal fittings so that it can be inserted into a pipeline. If the liquid in the pipe is perfectly clear, it can sometimes be difficult to tell if it is moving, but usually there are enough bits of sediment, small air bubbles, etc., in the liquid to give a clear indication of motion. Sometimes, small propellers or vanes are mounted in the sight glass and their motion indicates fluid flow.

The instruments described so far have been mainly of the type called *indicators* or *indicating instruments*. They have some kind of scale that enables an operator to obtain a reading. But often it is necessary to have a permanent record of instrument readings over a period of time; *recorders* are used for this purpose. These *recording instruments* make marks on a moving sheet of paper, usually in the form of a graph. The paper is often printed with lines that show the time the marks were made. Recording instruments frequently include a direct-reading, or indicating, device so that an operator can read the value at any instant. Such instruments are called indicator-recorders.

But instruments can do more than indicate or record measurements; they can also be used to control a process. Let us say there is a tank of water to be held at a certain temperature. If there is a temperature indicator on the tank an

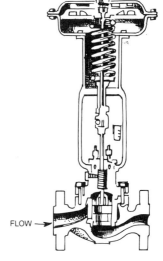

FLOW →

Control Valve

operator can see whether or not to open a steam valve to heat the tank or close a valve when the temperature is correct. It is fairly easy to devise an instrument that will automatically open the steam valve if the temperature is too low and close it when the proper temperature is attained. Such an instrument is called a *controller*, or *control instrument*. It may also indicate or record, or both. Thus we may have indicator-controllers, recorder-controllers or indicator-recorder-controllers. There are special types of valves designed to be operated by control instruments rather than by hand. They are called *control valves* and are very common in any automatically controlled plant.

In large continuous process plants, most of the plant's instruments are grouped in a special place called a *control room*. By using these instruments, an operator in the control room can tell what is happening anywhere in the plant. The operator can make adjustments in the process without leaving the control room, and can even shut down the whole plant if it should become necessary.

Discussion

1. What are the two most common instruments used for measuring temperature?

2. Explain how a thermocouple works.

3. What is a thermocouple well used for?

4. What is the name of the instrument used for measuring the temperature inside furnaces?

5. What is the most common kind of pressure-measuring device? Explain how it works.

6. How does a manometer work?

7. What is the simplest kind of flowmeter used in process plants? How does it work?

8. What problems are associated with orifice-plate flowmeters?

9. What is a rotameter? What else is it called? How does it work?

10. What is the purpose of using a sight glass?

11. What is an indicating instrument? Name some.

12. What are recording instruments used for?

13. What is a controller? Why are controllers important in the CPI?

14. What operates a control valve?

15. What would you find in a control room? What can be done from a control room?

Review

A. This unit described instruments for measuring pressure, temperature, and flow. From your own knowledge of chemistry, discuss why such measurements would be important in almost any chemical process.

B. Also from your own knowledge, make a list of some of the other properties or variables that might have to be measured in a chemical process plant.

C. Match the terms in the left column with the proper short definition in the right column. Only one definition is appropriate for each term.

1. Flowmeter   ____ Variable-area flowmeter

2. Recorder   ____ Transparent pipe section

3. Sight glass   ____ Any flow-measuring device

4. Thermocouple well   ____ Place from which a plant can be run

5. Control room   ____ Shows the value of a variable

6. Indicator   ____ Makes a record of values of variables

7. Thermometer   ____ Indicates pressure

8. Rotameter   ____ Works by expansion of liquid in a tube

9. Manometer   ____ Operated by a control instrument

10. Control valve   ____ Protects a thermocouple

UNIT SIX
FLUID FLOW

Special Terms

Fluid Dynamics: The science concerned with the flow of fluids—materials that flow, either liquids or gases.

Compressor: A machine used for pressurizing gases.

Centrifugal Pump: The most common type of pump. The rotating parts impart energy to the liquid, thereby increasing its pressure and velocity.

Impeller: The rotating part of a centrifugal pump. It consists of a disk (or pair of disks) with vanes.

Reciprocating Pump: A pump that consists of a piston moving back and forth (reciprocating) in a cylinder.

Duct: A tubular passage through which a fluid—usually a gas—is conducted. Ducts for gases are usually rectangular in cross section.

Density: The weight of a given volume of material. It is usually measured in terms of pounds per cubic foot or grams per cubic centimeter.

Viscosity: The resistance to flow of a fluid. It is usually measured in terms of the centipoise (cP) or the centistokes (cSt).

Throttling: The regulation of fluid flow by changing the size of the passage through which the material must flow.

Gate Valve: A common type of valve designed to be either fully open or fully closed.

Valve Trim: The moving parts of a valve that are made wet by the liquid flowing past.

Globe Valve: A common variety of valve designed for throttling fluid flow.

Vocabulary Practice

1. Define *fluid dynamics*. What are fluids?

2. What does a *compressor* do?

3. How does a *centrifugal pump* work?

4. What is the *impeller* of a centrifugal pump?

5. What is a *reciprocating pump*?

6. What is a *duct*? What is the usual shape of ducts used for gases?

7. What is meant by *density*?

8. What is the definition of *viscosity*? What units are used to express viscosity?

9. What is *throttling*?

10. What is a *gate valve*?

11. What parts of a valve are called the *trim*?

12. What is a *globe valve* used for?

Fluid Flow

In the chemical process industries, plants most often handle materials in a fluid state—that is, as a liquid or a gas. In general, fluids are much easier to transport from one place to another than are solids. Fluids can be transported in pipes in a continuous fashion, whereas solids-handling frequently requires rather complicated mechanical conveyors.

Since piping makes up so large a part of most process plants, the chemical engineer has to be thoroughly familiar with its design. The sizing of pipes is part of the subject of *fluid dynamics*, the science concerned with the flow of fluids. Fluid dynamics is also involved in the choice of pumps and *compressors*, both of which impart pressure energy to the fluid; it is this energy that causes flow. Pumps are used for pressurizing liquids and compressors for gases.

The most common type of pump used in the process industries is the *centrifugal pump*. In such a pump, a rotating wheel (or disk) with vanes on it, called an *impeller*, is used to impart energy to the liquid. The liquid enters the pump near the center of the impeller, where it is caught by the impeller's vanes. These impart a velocity to the liquid and centrifugal force carries it to the wall of the pump, where it exits. (The wall, or outer part of the pump, is called the casing.) Although the basic principle of the centrifugal pump is simple, there are many variations of pump design, depending on the kind of

Cross-section of a Centrifugal Pump

liquid to be handled. For example, centrifugal pumps may be designed for liquids containing large pieces of solid material (such as sewage), liquids at high temperatures, very abrasive materials (such as suspensions of sand in water), or very corrosive liquids (such as acids and alkalis). Centrifugal pumps are available in dozens of different materials, although the most common is carbon steel. They are supplied in all of the materials of construction described in Unit Three, as well as others. One of the most desirable characteristics of centrifugal pumps is that they provide a steady flow, without pulsations of pressure.

The second largest class of pumps used in process plants is the reciprocating type. All such pumps consist of pistons moving back and forth in cylinders. *Reciprocating pumps* can easily be designed to reach pressures far higher than those achieved by centrifugal pumps, and this is the area of their most common use. Unlike the centrifugal pump, the reciprocating pump imparts pressure in pulses—one at each stroke of the piston. One way of getting around this problem is by using a number of pistons in the pump, each of which reaches the end of its stroke (and, consequently, its highest pressure) at a different time. The pulsations of such a multiple-piston pump tend to average out so that the pulses are not severe enough to affect most processes.

Another use of reciprocating pumps is for metering liquids. Each stroke of the piston displaces the same volume of liquid, so that the output of the pump can be easily calculated. Although any

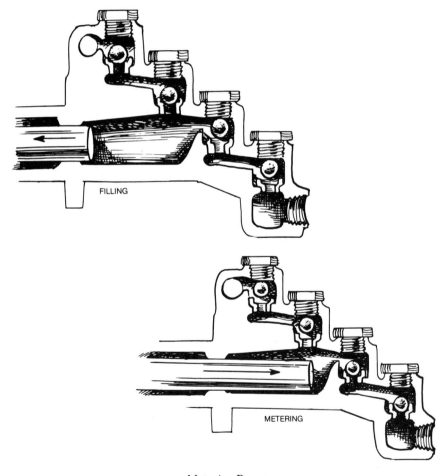

Metering Pump

reciprocating pump can be used in this way, some are especially designed for metering liquids. They are built so that the speed of the pump and stroke (distance travelled) of the piston are both adjustable. When constructed with several individually adjustable pistons, they are called proportioning pumps. (The use of proportioning pumps was described in Unit Two.)

Gases move in a pipe or *duct* whenever the inlet pressure is higher than the outlet pressure. Pressurization can be accomplished by means of either fans or compressors. Fans are used for low pressures.

Like pumps, compressors may be either of the centrifugal or reciprocating variety. The reciprocating compressor is very similar in principle to the reciprocating pump. (The common bicycle pump is actually a very simple variety of hand-powered reciprocating compressor.) Centrifugal compressors are generally of the turbine variety, and are similar in principle to the centrifugal pump, although far more complicated in actual design.

In the design and selection of both pumps and compressors, the *density* and *viscosity* of the fluid to be moved are of particular importance. The density of any material is the mass of a sample divided by its volume. It is usually expressed as grams per cubic centimeter or pounds per cubic foot. Viscosity is somewhat more difficult to define without using mathematical expressions, but it can generally be considered as resistance to flow. When poured from a jar, honey flows much more slowly than water. Thus honey has a higher viscosity (or is more viscous) than water. Both density and viscosity normally decrease with increasing temperature, although the changes in viscosity are more pronounced than the density changes. Some heavy fuel oils are so viscous at normal temperatures that they cannot be pumped. Consequently it is necessary to heat them so that the viscosity is reduced sufficiently to allow them to flow. Storage tanks for such fuel oils are equipped with steam coils for heating purposes.

Controlling the flow of liquids in pipes is done by means of valves. There are two basic classes of valves: on/off and *throttling*. An on/off valve, as the name implies, is either fully open (on) or fully closed (off). A throttling valve may be fully open, partially open (or, looking at it another way, partially closed), or fully closed. A partially open valve presents a resistance that slows down flow, so throttling valves are used to change the rate of flow of liquids. On/off valves are sometimes called stop or block valves because they either permit flow or stop it entirely.

The most common variety of valve used in process plants is called a *gate valve*. Its "gate" is a disk of metal that either closes off the pipe or is raised up into the body of the valve so that the opening in the pipe is unobstructed. When the gate is in the raised (or open) position, the body of the valve acts like a part of the pipe and the fluid flows almost as if the valve were not there. It is possible to use a gate valve for throttling by closing the gate only part of the way, but the valve is not designed for such service and will often be damaged if it is used in this way. The liquid flow tends to wear away the sharp edges of the gate so that it will not close tightly and will leak. When used properly, the gate in the open position is withdrawn into the body of the valve so there is no wear on it from liquid flow.

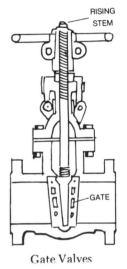

Gate Valves

A gate valve is usually opened or closed by turning a handwheel. In the more common pattern of gate valve, a threaded rod (or stem) is attached to the gate and is moved by the handwheel. When the gate is open, the stem projects above the handwheel and is therefore called a rising-stem valve. Rising-stem valves are preferred because, by looking at the stem, it is easy to see whether the valve is open or closed. When designing a piping system, space must be provided above the valve so the stem will not hit anything when it rises. If this space cannot be provided, the engineer can specify a nonrising-stem valve. These can sometimes lead to dangerous situations because an opera-

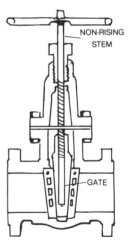

tor may forget that it is a nonrising-stem valve and assume it is closed—because he cannot see the stem—when the valve is actually open. Nonrising-stem gate valves should be painted, or otherwise marked in some distinctive manner, to prevent confusion. Gate valves are available for every size of pipe and in a wide variety of construction materials. The *valve trim* (those moving parts that are made wet by the fluid in the pipe) is frequently made of material that is harder and more corrosion-resistant than the remainder of the valve.

When it is necessary to control the quantity of liquid flowing in a pipe (that is, to throttle it), the most common valve used is the *globe valve*. To pass through a globe valve, liquid must turn through two right angles. This causes swirls and eddies in the liquid stream, which wastes energy. This is in contrast to the "straight-through" flow pattern in a gate valve. This waste of energy in globe valves shows up as a pressure drop in the pipeline and in increased pumping costs. Although a globe valve can be used for on/off service without damage to the valve, it is normally restricted to throttling service because of its energy losses.

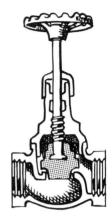

Globe Valve

Globe valves are usually installed so that the flow is upward around the shut-off disk. This is to prevent the accumulation of solids that might prevent full closure above the disk. It also permits maintenance to be performed on the upper part of the valve when it is closed but still in the pipeline, because this part is isolated from the high pressure (input) side of the line. To facilitate proper installation, globe valves usually have an arrow cast, or otherwise marked, on the body of the valve. The valve should be installed so that the arrow points in the direction of liquid flow. Globe valves are available in a wide size range, although very large globe valves are less common than very large gate valves. They are also available in most construction materials, both for the body of the valve and for the trim.

As described in the preceding unit on instrumentation, some valves are automatically opened and closed by an instrument rather than by hand. These control valves are generally a variety of globe valve, since their most common use is for throttling. A special and rather complex mechanism is needed to accomplish automatic control so these valves are much larger and far more expensive than the common hand-operated varieties.

Discussion

1. Why must a chemical engineer have a good knowledge of fluid dynamics?

2. Describe the way a centrifugal pump operates.

3. What are some of the special services for which centrifugal pumps are designed?

4. What kinds of materials are used for making centrifugal pumps?

5. What is one desirable characteristic of a centrifugal pump?

6. How does a reciprocating pump operate?

7. What is the most common use for reciprocating pumps?

8. What is an advantage of reciprocating pumps with a number of pistons?

9. Why are reciprocating pumps used for metering liquids?

10. Under what conditions will a gas move in a duct?

11. What are the two varieties of compressors?

12. How would you determine the density of a fluid?

13. Which is more viscous, honey or water?

14. How does the viscosity of a fluid change as the temperature rises?

15. What are the two main classes of valves?

16. What is the difference between an on/off valve and a throttling valve?

17. What is another name for an on/off valve?

18. Is a gate valve designed for use in throttling? Explain why.

19. What is a rising-stem gate valve?

20. When would an engineer specify a nonrising-stem gate valve? What kind of problems might this cause?

21. Are the same materials always used for a valve body and the valve trim?

22. What kind of valve is usually used for throttling service?

23. Can globe valves be used as stop valves? Are they usually used this way? Why?

24. How might an engineer check to see that a globe valve is installed in the right direction?

Review

A. So far in this book, a number of things related to plant safety have been mentioned. They include pressure vessel codes, dikes in tank farms, the use of rising-stem gate valves, and emergency shutdown procedures. Discuss how all of the factors mentioned in the text—and any others that you may also know of—contribute to the good safety record of the chemical process industries.

B. Indicate which of the following statements are true and which are false.

1. An impeller is an important part of a reciprocating pump.

2. A gate valve is a common type of on/off valve.

3. The density of a liquid changes as the temperature changes.

4. The viscosity of a liquid changes as the temperature changes.

5. Density is another name for viscosity.

6. Globe valves may be used for throttling, but this is not recommended because parts of the valve may be damaged.

7. Fans are used when only low pressures must be imparted to gases.

8. All centrifugal pumps are made of carbon steel.

9. A bicycle pump is really a kind of compressor.

10. Liquid enters a centrifugal pump near the outer wall, or casing.

11. Fluid dynamics is the science concerned with the flow of heat.

12. Compressors are made in both rotary and reciprocating types.

13. Reciprocating pumps are generally used when high pressures must be achieved.

14. Centrifugal pumps are commonly used for metering liquids.

15. Valve bodies and valve trim are often made of different materials.

16. A globe valve usually has an arrow on its body.

17. The impeller of a centrifugal pump rotates while the pump is operating.

18. Gate valves are available for use only with small sizes of pipe.

19. Some centrifugal pumps are designed to pump corrosive liquids.

20. A chemical engineer needs to understand fluid dynamics.

21. A proportioning pump is a kind of metering pump.

22. Some fuel oils are very viscous.

23. A problem of centrifugal pumps is that the flow from them pulsates badly.

24. Liquids and gases are both fluids.

25. Gate valves cause greater energy losses in a pipeline than do globe valves.

UNIT SEVEN
HEAT FLOW (THERMODYNAMICS)

Special Terms

Thermodynamics: The science dealing with the generation and flow of heat, and the relationship between heat and other forms of energy (chemical, mechanical, and electrical).

Sparger: A perforated pipe or other container used for releasing gas in the form of small bubbles into liquid.

Heat Exchanger: A device used to transfer heat from one fluid to another.

Shell-and-Tube Heat Exchanger: The most common variety of heat exchanger used in process plants. It consists of a number of tubes inside a cylindrical vessel called a shell. One fluid flows inside the tubes, the other outside the tubes but within the shell.

Tubeside: Within the tubes of a shell-and-tube heat exchanger.

Shellside: Outside the tubes, but within the shell, of a shell-and-tube heat exchanger.

Steam Trap: A device that automatically permits liquids to drain from a line, but prevents the exit of steam.

Condensate: The material formed when vapor is cooled and changed, or condensed, into liquid.

Dowtherm: A specialized liquid with a high boiling point. It is used in a heat exchanger to transfer heat to other materials.

Nozzle: An opening in a vessel to which a pipe can be connected; a part which directs the flow of liquid when connected to a hose.

Countercurrent: A kind of flow in which two liquids move in directions opposite to each other. When both move in the same direction, the flow is said to be *cocurrent*.

Condenser: A heat exchanger used for cooling vapor and condensing it into liquid.

Laminar Flow: The smooth flow in a pipe in which the fluid in the pipe's center moves at the highest speed and the fluid near the pipe's walls moves at the lowest speed.

Turbulent Flow: A flow that is agitated, rather than smooth, and contains swirls and eddies.

Reynolds Number: A mathematical technique for predicting the transition between laminar and turbulent flow in a pipe.

Vocabulary Practice

1. Define *thermodynamics.*

2. What is a *sparger?*

3. What is a *heat exchanger* used for?

4. Describe a *shell-and-tube heat exchanger.*

5. What does *tubeside* mean?

6. What is the *shellside* of a heat exchanger?

7. What does a *steam trap* do?

8. What is *condensate?*

9. What is *Dowtherm?* What is it used for?

10. What is a *nozzle?*

11. What is meant by *countercurrent?* What is its opposite?

12. What is a *condenser* used for?

13. Describe *laminar flow.*

14. How does *turbulent flow* differ from laminar flow?

15. What is the *Reynolds number* used for?

Heat Flow (Thermodynamics)

In almost any chemical process plant, materials are constantly being heated or cooled. This is done for a variety of reasons. As we have seen in the previous unit, it is often necessary to heat viscous materials in order to make them flow easily. Another reason for heating is to melt solids, or to change liquids into vapors by boiling them. Chemicals may be heated to increase the speed of a chemical reaction or to make it occur at all. A handy rule known to most chemical engineers is that a 10°C increase in temperature will approximately double the rate of most reactions. But some chemical reactions give off heat as they occur; such materials generally require cooling if the reaction is to be properly controlled. Cooling is also required to change, or condense, vapors into liquids and to freeze liquids into solids.

In many reactions the temperature influences the products being formed. Consequently the temperature must be controlled so that the materials remain at the temperature that produces the highest yield of the most desired product.

The science concerned with the generation and transfer of heat is called *thermodynamics*; this subject is emphasized in every chemical engineer's education. In almost all chemical process plants, heat is generated by the burning of fossil fuels—coal, oil, or natural gas. The heat given off by the burning fuel may be used directly, but most often it is used to generate steam which is piped to the place where heat is needed.

The easiest way of heating a liquid with steam is to bubble the steam directly into the liquid. For example, steam can be supplied to a pipe immersed in the liquid; a number of small holes are drilled in the pipe to allow the steam to escape in small bubbles. Such a device is called a *sparger*. The bubbles of steam rise up, transferring heat to the liquid and mixing it at the same time. During this process, some or all of the steam condenses into water, which remains in the liquid being heated. Consequently, this kind of heating, called direct injection of steam, can be used only if the presence of condensed water is not a problem.

In most cases, direct injection of steam is impossible; the steam and the liquid to be heated must be kept separate. To heat the liquid without direct contact, a piece of equipment called a *heat exchanger* is used; it has a barrier of some solid material between the heating medium and the material to be heated. Many varieties have been invented, but by far the most common type is the *shell-and-tube heat exchanger.*

The material to be heated may be *tubeside* (passed through the tubes), with the steam *shellside* (outside the tubes). It is possible to have steam on the tubeside and the other material on the shellside, but this is uncommon. Usually the steam is on the shellside so that it condenses on the outside of the tubes. The resulting water falls to the bottom of the shell where it is

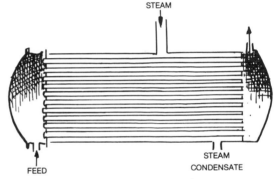

Shell-and-Tube Heat Exchanger

allowed to pass out of the heat exchanger by means of a special device called a *steam trap*. This is designed to permit the escape of *condensate* (water condensed from the steam) while trapping the steam (preventing it from escaping).

At normal atmospheric pressure, water boils at 212°F (100°C) and its steam is at the same temperature. If the pressure is increased, the boiling point of the water (and the temperature of the steam being produced) rises. To use high-temperature steam, it is necessary to design and build all equipment to withstand the high pressures associated with the high temperatures. High-pressure equipment is expensive. Sometimes, to provide high temperatures without having to deal with high pressures, chemical engineers use special heat-transfer liquids for heating. These are materials that remain liquid at much higher temperatures than water. Two common heat-transfer liquids are known as Dowtherm A and Dowtherm E. (*Dowtherm* is a trademark of the Dow Chemical Company, which manufactures and markets the products). Dowtherm

A is a mixture of two chemicals: diphenyl and diphenyl oxide. It boils at about 450°F (230°C). Dowtherm E is o-dichlorobenzene, which boils at about 300°F (130°C). Some mineral oils are also sold for use as heat-transfer liquids at temperatures ranging to about 600°F (315°C), but most tend to deteriorate if kept at high temperatures for a long time. (The Dowtherm materials are very stable, even after many years of use at high temperatures.)

If a material is to be cooled, a shell-and-tube heat exchanger can also be used. The cooling medium is normally water, chilled water, or cold brine. As mentioned earlier, brine (a solution of salt in water) is used because it freezes at a lower temperature than pure water. Sometimes liquids are cooled by circulating them through pipes exposed to the atmosphere; this is called an air-cooled heat exchanger. The cooling process is speeded if a fan is used to blow cool air rapidly over the tubes.

When a shell-and-tube heat exchanger is used for cooling, the material to be cooled may be either on the tubeside or the shellside, with the cooling medium on the other side. If one of the two liquids is more corrosive than the other (brine, for example, can be corrosive) it is usually on the tubeside. The tubes can then be made of a corrosion-resistant material, with a cheaper material for the shell. If a corrosive material is used on the shellside—in contact with both the inner surface of the shell and the outer surface of the tubes— both the tubes and shell must be able to withstand corrosion.

It is possible to operate shell-and-tube heat exchangers in either the horizontal or vertical position, but horizontal exchangers are most common. When steam is used for heating, it is always introduced through the upper opening, or *nozzle*, in the shell, so that the lower opening can be used for draining condensate. If a pool of condensate builds up in such an exchanger (usually due to a malfunctioning or wrongly sized steam trap) and submerges some of the tubes, the heat transfer suffers because condensing steam transfers heat more rapidly than does hot water. When a liquid is used on the shellside, it is always introduced through the lower nozzle and withdrawn through the upper nozzle. In this way the engineer can be assured that the exchanger remains full, with liquid surrounding all the tubes.

Shell-and-tube heat exchangers are so widely used that they are available as stock items from many manufacturers. There are a number of standard designs used throughout the industry. These are detailed in a publication of the Tubular Exchanger Manufacturers Association (TEMA) and are therefore called the TEMA Standards. However, no group of standard exchangers could fulfill all

Shell-and-Tube Heat Exchanger

needs, so many exchangers are custom designed and manufactured. The size of the shell-and-tube heat exchanger needed for any particular job depends mainly on the amount of heat to be transferred and the volume of fluid to be handled. The calculations are learned in college by most chemical engineers, but computer programs now do them automatically.

Although most heating is done with steam or heat-transfer liquids such as Dowtherm, other liquids may also be used as the heating medium. This is most commonly done when there is one liquid stream in a plant that needs to be heated and another that needs to be cooled. Passing both streams through the same heat exchanger will accomplish both objectives simultaneously. When a heat exchanger is used with liquids on both the tubeside and the shellside, the two liquids are usually made to flow in opposite directions; this is known as *counter-current* flow. If both materials flow in the same direction the flow is called *cocurrent.*

Sometimes a heat exchanger is designed to chill the vapors of a substance and condense it into a liquid. A special exchanger of this kind is called a *condenser* because of the way it is used, although it is not too different from any other exchanger. When used as condensers, shell-and-tube heat exchangers are often mounted vertically.

Vertical
Heat Exchanger

When a fluid flows in a pipe, it moves fastest at the center of the pipe and slowest at the wall because of friction between the pipe and the flowing fluid. When fluid flows relatively slowly, there can be a fairly thick layer of fluid near the pipe wall that hardly moves at all. This condition is known as *laminar flow*, and makes for poor heat transfer because the static layer of fluid near the wall acts as a resistance to the movement of heat. When fluid is flowing rapidly, swirls and eddies are set up that reduce the thickness of the static layer near the pipe wall, and thereby decrease the resistance to the movement of heat—or increase the rate of heat transfer. This kind of agitated flow is known as *turbulent flow*.

The speed at which laminar flow changes to turbulent flow was investigated early in this century by an experimenter named Osborne Reynolds. He worked out a relationship, known as the *Reynolds number*, that helps to determine the transition from laminar to turbulent flow. The Reynolds number is found by multiplying the inside diameter of the pipe by the fluid velocity and the fluid density, and then dividing by the fluid viscosity. The units used for the variables are selected so that all the units cancel out; hence, the Reynolds number is known as a dimensionless number. (For example, diameter in centimeters, velocity in centimeters per second, density in grams per cubic centimeter and viscosity in grams per centimeter-second.)

When the Reynolds number is less than 2,100, flow is always laminar. When it is above 4,000, flow is always turbulent. Between Reynolds numbers of 2,100 and 4,000, the flow may be either laminar or turbulent, depending on conditions at the entrance of the pipe and on the distance downstream from the entrance.

Discussion

1. What are some of the reasons for heating materials in chemical process plants?

2. What are some reasons for cooling materials?

3. What is the simple rule that relates temperature to rate of reaction of chemicals?

4. How is heat usually obtained in process plants?

5. Describe heating by direct injection of steam.

6. What might be used to bubble the steam into a liquid?

7. What is a disadvantage of direct injection of steam?

8. What is a heat exchanger used for?

9. What is the most common type of heat exchanger?

10. What is the difference between the tubeside and the shellside of a shell-and-tube heat exchanger?

11. How is condensed steam usually removed from a heat exchanger?

12. At what temperature does water boil under atmospheric pressure? What is the temperature of the steam produced?

13. How can the temperature of steam be increased? What is the disadvantage of doing this?

14. What materials, besides steam, are used for heating purposes?

15. What liquid is usually used for cooling in shell-and-tube heat exchangers?

16. Describe an air-cooled heat exchanger.

17. Why is it desirable to use a corrosive material on the tubeside of a tubular exchanger rather than on the shellside?

18. Are shell-and-tube exchangers usually mounted in the horizontal or vertical position?

19. Is steam usually introduced through the upper or lower nozzle of an exchanger's shell? Why?

20. Is cooling liquid normally introduced through the upper or lower nozzle of a shell-and-tube heat exchanger? Why?

21. What would you call a heat exchanger that is designed to condense a vapor?

22. Describe the difference between laminar and turbulent flow.

23. What is the name of the relationship that enables an engineer to predict whether flow is likely to be laminar or turbulent?

Review

A. Choose any chemical process with which you are familiar and discuss at what stages of the process materials have to be heated and at what stages they have to be cooled.

B. Complete the following sentences with the proper word or phrase.

1. In a shell-and-tube heat exchanger, liquid outside the tubes is said to be on the _____.

2. A _____ is an opening on a vessel to which a pipe can be connected.

3. When fluids are flowing _____, they are moving in opposite directions.

4. A _____ automatically permits condensate to leave but does not permit steam to leave.

5. If the flow in a pipe is _____, heat transfer will be better.

6. A _____ may be used to bubble gas into a liquid.

7. A material made by the Dow Chemical Company, called
_____, is used as a heat transfer liquid.

8. An engineer studies _____ to learn about the flow
of heat.

9. The _____ is used to predict the transition between
laminar and turbulent flow.

10. The _____ heat exchanger is the most common type
used in process plants.

11. A _____ is a heat exchanger used to change vapor
into liquid.

UNIT EIGHT
SEPARATING BY HEATING

Special Terms

Distillation: The separation of two or more liquids, with different boiling points, by evaporating them and condensing the resulting vapors.

Binary Distillation: The distillation of a mixture of two liquids.

Still: A piece of equipment used for distillation.

Distillation Column: A modern variety of a still that can be used for distillation of liquids as a continuous process.

Reboiler: A heat exchanger that uses either steam or hot furnace gases to revaporize material reaching the bottom of a distillation column.

Distillation Tray (Plate): A part of a distillation column through which rising vapor is intimately mixed with descending liquid.

Bubblecap Tray: A type of distillation tray.

Overhead Product (Overheads): The part of the condensed vapor, from the top of a distillation column, that is withdrawn from the process.

Bottom Product (Bottoms): The material withdrawn from the bottom of a distillation column.

Multicomponent Distillation: The distillation of a mixture of more than two substances. Distillation of petroleum is a common example of this process.

Reflux: The condensed vapor, from the top of a distillation column, that is returned to trickle down the column.

Reflux Ratio: The ratio between the amount of condensed vapor returned to the column as reflux and the amount withdrawn as overhead product.

Evaporator: A device for concentrating solutions by removing part of the liquid as a vapor. The process is called *evaporation*.

Multiple-effect Evaporator: A type of evaporator in which vapor from one stage of the system is used to heat liquid in another stage.

Vocabulary Practice

1. What is *distillation*?

2. How many different materials are separated in *binary distillation*?

3. What is a *still*?

4. What is a *distillation column*?

5. What is a *reboiler*?

6. What part of a distillation column is a *distillation tray*?

7. What is a *bubblecap tray*?

8. What is *overhead product*? What else is it called?

9. What is *bottom product*? What else is it called?

10. Define *multicomponent distillation*. Give an example.

11. What is *reflux*?

12. What is the *reflux ratio* in a distillation column?

13. What is an *evaporator* used for?

14. What is a *multiple-effect evaporator*?

Separating by Heating

The chemical engineer often needs to separate mixtures of materials. These mixtures sometimes occur naturally; petroleum, for example, is a mixture of a great many chemicals. Some mixtures occur as a result of chemical reactions; few reactions will produce a single pure chemical. Often the desired material is produced with some that are undesirable; the resulting mixture must be separated so as to recover the wanted one in as pure a state as possible.

Distillation has been used to separate mixtures of liquids since the earliest days of chemistry. It is based on the principle that if a mixture of liquids is heated, some of the ingredients will evaporate faster than others and this property can be used to effect a separation. Let us imagine that we have a mixture of two liquids, A and B, in equal parts—the mixture contains 50% of Liquid A and 50% of Liquid B. If we heat the mixture until it boils, we may find that the vapor contains 75% of Liquid A and only 25% of Liquid B. Liquid A is evaporating faster than Liquid B.

In distillation, a liquid mixture is heated until it vaporizes, and the vapor is then condensed back into a liquid. Two liquids are generally quite easy to separate by distillation if the pure materials have boiling points that differ by a considerable amount, but modern techniques permit the separation of liquids whose boiling points are close together. Separation of mixtures of two liquids is called *binary distillation.*

Distillation can be carried out as a batch process; this was the original kind of distillation and is still sometimes done today. For example, alcoholic beverages, such as brandy and some whiskies, are distilled by batch processes. (Essentially, these distillations consist of separating ethyl alcohol from water; consequently, they are binary distillations.) However, continuous distillation is faster and more efficient than the batch process, and most modern *stills* are of the continuous type. Continuous stills are called *distillation columns* because they look like circular architectural columns or large-diameter pipes standing vertically on one end. But the interior of a pipe is empty space, and a distillation column is anything but empty.

Heat is supplied to a distillation column by a heat exchanger called a *reboiler*. Here liquid is vaporized and starts to ascend the column. At the top of the column, vapor is condensed in another heat exchanger (a condenser) and reintroduced into the column, where it starts to trickle downwards. Spaced at intervals throughout the column are flat *plates*, usually called *trays*, that are designed to ensure thorough mixing between the vapor that is ascending the

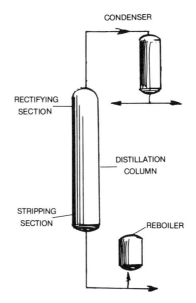

Distillation Column

column and the liquid that is descending. It is particularly easy to understand the process taking place on a *bubblecap tray* (although there are many that are simpler to build and more efficient).

As shown in the diagram, the tray is covered with a layer of liquid. The lower edges of the bubble caps are below the surface of the liquid. Vapor rising in the column comes up under the bubble cap and is forced to bubble through the pool of liquid on the tray, thereby thoroughly mixing the vapor and the liquid. The liquid on the tray is cooler than the rising vapor; as the vapor bubbles through the liquid, any material in the vapor that is close to its condensation point (a temperature identical to its boiling point) will tend to condense and remain in the pool of liquid. At the same time the vapor, in cooling, is heating the liquid. Any component in the liquid that is near its boiling point tends to vaporize and be carried up the column.

A temperature gradient occurs along the height of the column, with the highest temperature at the bottom near the reboiler and the lowest at the top of the column near the condenser. After the column has been in operation for a time, the components of the mixture being distilled tend to distribute themselves on the various trays of the column according to their boiling points—the materials with the lowest boiling points will be near the top.

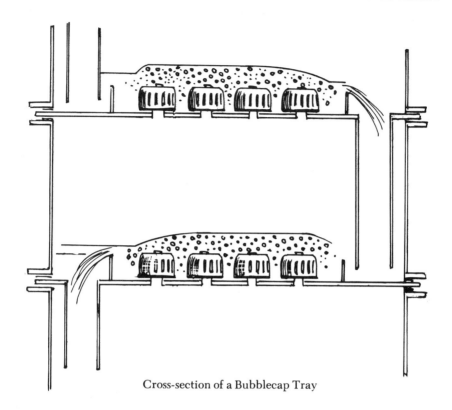

Cross-section of a Bubblecap Tray

Bubblecap Tray

Once the column is working, feed can be continuously introduced (usually near the middle) and product withdrawn from the top of the column, from the bottom, and at various intermediate points. If the material being distilled contains only two components (such as ethyl alcohol and water), all of the desired product (the alcohol) will be taken from the top of the column. This is called *overhead product*, or simply *overheads*. The water, which is unwanted, is taken from the bottom of the still and discarded. This is called *bottom product*, or *bottoms*. If the material being distilled contains many components (as is the case with petroleum), products of desired composition may be withdrawn from intermediate trays, as well as from the top and bottom of the still. This is known as *multicomponent distillation*. The product withdrawn from intermediate trays may not be pure; it may be a mixture of several liquids with boiling points close together. Note that in petroleum distillation the bottoms are not thrown away (as was the case in the water/alcohol distillation) but are themselves valuable products. When the product withdrawn from one of the plates of a still is a mixture, it may be salable as such —kerosene, gasoline, and benzine

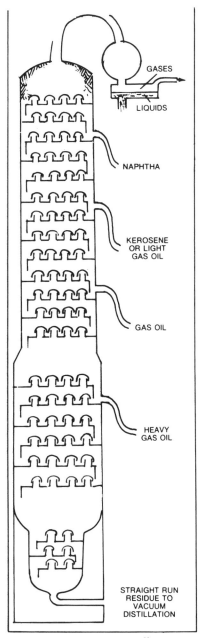

Multicomponent Distillation

are such mixtures. However, if desired, the mixtures may be further separated in another still.

The liquid returned to the distillation from the overhead condenser is known as *reflux*. In general, the more overheads returned to the column as reflux, the better the separation between the components of the feed. The ratio between the amount of overheads taken as product and the amount of condensate returned as reflux is known as the *reflux ratio*.

Columns can be designed to operate at high or low reflux ratios. Obviously, a low reflux ratio is desirable because more of the condensate can be taken as product, but a column designed for this must be quite tall with a large number of trays. It will therefore be very expensive. If a high reflux ratio is used, the column can be shorter and less expensive. However, high reflux means that more heat must be introduced into the reboiler to revaporize the reflux. Thus, although the initial cost of the column is lower, the operating cost—for extra fuel—will be higher. It is the job of the design engineer to balance the high initial cost of the tall column against the higher operating cost of the shorter one and arrive at a compromise that will have the lowest total cost over the life of the column.

Another kind of separation that uses heat is the concentration of a solution of some material, often a salt, in liquid. Most often, the solvent is water and the concentration is accomplished by boiling the solution so that the water turns into a vapor and is removed. The equipment in which the solution is boiled and concentrated is known as an *evaporator*. In *evaporation*, what remains after the liquid is boiled off is the desired product. The vapor has no value and is discarded. In distillation, the vapor is usually a valued product. Another difference between distillation and evaporation is that there is never any attempt to separate the components of the vapor.

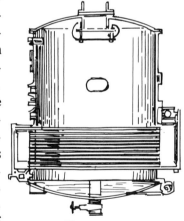

Evaporator

An evaporator is a special kind of heat exchanger, usually vertical. Heating is usually done with low-pressure steam. The vapor leaving the evaporator may be condensed, creating a vacuum in the evaporator. The effect of the vacuum is to make the liquid in the evaporator boil faster and at a lower temperature. Sugar solutions are usually concentrated in such evaporators.

Although the vapor from an evaporator is an unwanted material, it does have some value because it contains heat. It is possible to use the hot vapor to heat the contents of a second evaporator. Sometimes vapor from the second evaporator can be used to heat a third one, and so on. These series of evaporators are known as *multiple-effect evaporators*, and have the advantage of making more efficient use of the heat in the steam than is possible with a single evaporator. Multiple-effect evaporators with as many as four stages are common.

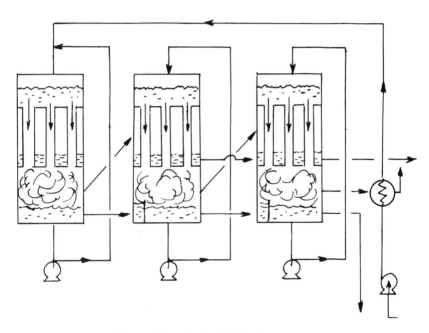

Three-effect Vertical Tube Evaporator

Discussion

1. Why do chemical engineers need to separate mixtures of liquids?

2. What is the principle on which distillation is based?

3. Is distillation a batch or a continuous process?

4. Give an example of binary distillation.

5. Are distillation columns continuous stills?

6. How is heat supplied to a distillation column?

7. What is used to condense vapors from a distillation column?

8. What is the purpose of a distillation column tray?

9. Name one kind of distillation tray. Describe it.

10. Describe that process that goes on at each distillation tray.

11. From what part of the column are overheads taken?

12. Are products withdrawn only from the top or bottom of a column?

13. Are the products taken from intermediate trays of the column always single pure materials?

14. Are bottoms always waste products?

15. Where does reflux come from?

16. Is a high reflux ratio preferable to a low one? Why?

17. What is a disadvantage of designing a column to use a low reflux ratio?

18. What are some differences between distillation and evaporation?

19. How are evaporators usually heated?

20. Is there any way to use the vapor from an evaporator?

21. What is the advantage of a multiple-effect evaporator?

Review

A. Cooks often boil soups and sauces to make them more concentrated. Is this anything like the process of evaporation used in chemical process plants? Explain the reasons for your answer.

B. Indicate which of the following statements are true and which are false.

1. Some mixtures of chemicals occur naturally.

2. Separating a mixture of three liquids is called binary distillation.

3. The bubblecap tray is the most efficient distillation tray ever invented.

4. The materials withdrawn from a distillation column are always pure substances.

5. A low reflux ratio makes a distillation column cheaper to operate.

6. In both distillation and evaporation, a liquid is changed into a vapor.

7. Multiple-effect evaporators are less efficient than single evaporators.

8. The vapor from a multiple-effect evaporator is called reflux.

9. Multicomponent distillation is used in processing petroleum.

10. Sugar solutions are often evaporated with the help of a vacuum.

11. The liquid to be distilled is always fed into a distillation column at the top.

12. Sometimes the product of one distillation column is redistilled in another column.

13. While a still is operating, a bubblecap tray is covered with a layer of liquid.

14. Heat is supplied to a distillation through a heat exchanger called a condenser.

15. When vapor bubbles through the liquid on a distillation tray, some of the vapor is condensed and some of the liquid vaporizes.

16. Distillation can be carried out either as a continuous process or as a batch process.

UNIT NINE
SEPARATING WITHOUT HEATING

Special Terms

Size Classification: The process of separating granular or powdered materials into batches of large, intermediate, and small particles.

Grizzly: A size-classification device for large lumps of solids. It consists of inclined metal bars with spaces between them. Any material smaller than the space between the bars will fall through; larger sizes slide to the end of the grizzly where they are collected.

Sieve: A size-classification device made of wire cloth or perforated sheet metal. When the sieve is shaken, small particles fall through the openings and larger particles remain on top. Sieves made of wire cloth are commonly called *screens*.

Sedimentation (Settling); A process in which particles of solids suspended in a liquid are separated by allowing them to settle by the force of gravity.

Thickener: A piece of equipment in which sedimentation is carried out as a continuous process.

Filtration: A process by which solid particles may be separated from a liquid by forcing the liquid through a woven cloth with openings too fine to permit the particles to pass.

Filter Cloth: The special finely woven cloth used for filtration.

Filter Press: A structure on which metal or plastic plates and frames are suspended and pressed together. Filter cloth is placed between the plates and frames; the material to be filtered is forced through the cloth under pressure.

Filtrate: The clear liquid that flows from a filter after the solids have been removed.

Filter Cake (Cake): The solids removed during filtration.

Continuous Filter: A machine for carrying out filtration as a continuous process.

Centrifuge: A machine for separating solids from liquids or for separating two liquids of different densities which cannot mix. The separation is accomplished by the use of *centrifugal force* created by rapid whirling.

Cyclone Separator: A device for separating solid particles from a gas. The cyclone causes the gas to swirl, thereby generating a centrifugal force that causes the particles to move to the wall of the cyclone.

Gas Scrubber: A device that separates solid particles from a gas or one gas from another. It uses a spray of liquid (usually water) that traps the particles or dissolves a gas.

Vocabulary Practice

1. What is meant by *size classification*?

2. Describe a *grizzly*.

3. What is a *sieve*? How is it used? What are wire cloth sieves sometimes called?

4. What is *sedimentation* used for?

5. What does a *thickener* do?

6. Describe the process of *filtration*.

7. What is a *filter cloth*?

8. What is a *filter press*?

9. What is *filtrate*?

10. What is *filter cake*?

11. What is a *continuous filter?*

12. What does a *centrifuge* do?

13. What is a *cyclone separator* used for?

14. What two things is a *gas scrubber* used for? How does it work?

Separating Without Heating

In addition to those processes that use heat for separation, there are a number of processes that do not require heat. Perhaps the simplest of these is *size clas-sification*, which involves sep-arating large particles, or lumps, of solids from small ones. Large lumps of coal or ore, for example, can be separated from smaller ones by letting the mate-rial slide down a chute com-posed of parallel bars of metal. Small lumps fall between the bars into a container; larger

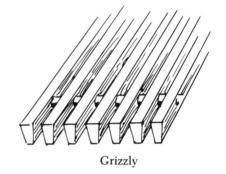

Grizzly

ones slide to the end of the chute where they, too, are collected. Such a device is called a *grizzly.* It may be shaken to help move the lumps along, and several may be used in series—each having a wider space between the bars—to separate the lumps into a number of size groups.

Grizzlies are used only for large pieces of material. For small particles it is more common to use *sieves* made of perforated sheet metal, or a type of sieve made of wire cloth, called a *screen.* Mate-rials may be passed through a series of sieves, each having holes smaller than the last. In this way, the material is separated into a number of size classes called sieve fractions; each fraction is small enough to pass through the next larger sieve, but too large to pass through the one on which it is finally retained.

This procedure is sometimes followed in the laboratory: a weighed quantity of granular or powdered material is passed through a series of vibrating screens; the fraction remaining on each screen is weighed and calculated as a percentage of the original weight. Such a procedure is known as a size analysis or a sieve analysis. Standards have been set up for the series of sieves used so that the results from one laboratory will be comparable to the results from any other laboratory.

Another kind of separation that does not use heat is the removal of suspended solids from a liquid. If the solids have a higher density than the liquid (as is usually the case) they may be removed by simply allowing the liquid to remain in a tank for a period of time. During this time, the force of gravity will settle the solids on the bottom; this is called *sedimentation*, or settling. There are continuous sedimentation devices in which solids are continuously removed from the bottom of the tank and clear liquid from the top; these are called *thickeners*. The solids removed from a sedimentation device always contain some liquid; they are watery or muddy and require further treatment if all the moisture is to be removed.

Another heatless method of removing solids from liquid is by a process similar to sieving, called *filtration*. In this process, the solids-bearing liquid is forced through a material (usually a finely woven cloth) that allows the liquid to pass through but holds back the solids. As a layer of solids builds up on the surface of the *filter cloth*, the liquid must first pass through this solids layer. Since the openings through the solids layer are usually smaller than the openings in the filter cloth, such a filter will hold back finer particles after it has been in use for a time than when it is clean. Consequently, the first liquid to pass through the filter may be saved and passed through again (recycled) after the solids layer has built up. Another way of accomplishing the same thing is to first coat the filter cloth with a layer of material that will provide finer openings than the filter cloth. These materials are called filteraids, and the initial layer is called a precoat.

Much filtration is done in devices called *filter presses*. The most common is called a plate-and-frame filter press because it consists of alternating perforated plates and open sections called frames. Filter cloth is placed over both sides of each plate and the assembly of plates and frames is squeezed together in the press. Holes in the plates and frames act as pipes to distribute the liquid to be filtered into the frame sections, where it passes through the filter cloth into the perforated plate sections, and then out through a dis-

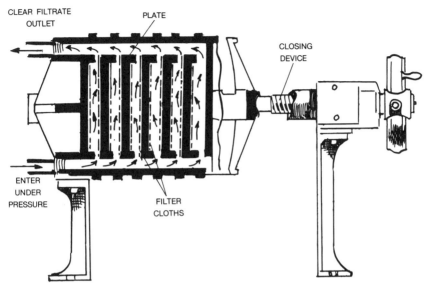

CLEAR FILTRATE
OUTLET

PLATE

CLOSING
DEVICE

ENTER
UNDER
PRESSURE

FILTER
CLOTHS

Filter Press

charge opening. The clear liquid discharged is called the *filtrate*, and the solids trapped in the frames on the surface of the filter cloth are called the *cake* or *filter cake*. Depending on the process, the desired product may be the clear filtrate or the cake; the other material is usually waste. However, in a few filtrations both the filtrate and cake are valuable products.

A filter press is normally run until the frames are full of cake; then it is shut down for cleaning. This makes it a batch process. Filter presses can be used in a continuous process plant by providing at least two, so that one can operate while the other is being cleaned. Cleaning is done by opening the press and shaking the frames so that the cake falls into a trough below the filter press. In some filtrations, the cloths must be replaced at each cleaning cycle; in others they may be used for many cycles. The filter press is then closed and is ready for another cycle. If filteraid is used as a precoat, a watery mixture of it will first be made up and pumped through the press until the required thickness is deposited on the filter cloths.

Although filter presses are very common, they have one major disadvantage: cleaning them is a tedious process that requires much hand labor, which can be very expensive. However, they are relatively inexpensive to buy and maintain—there is not much that can go wrong with a filter press.

Continuous filters are also available, but they are much more complex machines. This makes them much more expensive to buy and to maintain. Their advantage is that they require very little labor during operation. The most common continuous filter is the rotary drum type, which consists of a revolving drum covered with filter cloth. Considerable pressure is needed to force the liquid through the filter

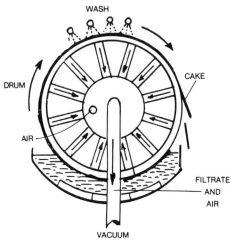

Rotary Drum Filter

cloth, so the entire filter may be enclosed in a pressure vessel.

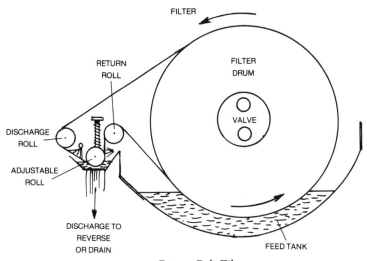

Rotary Belt Filter

Otherwise, the interior of the drum may be evacuated (put under a vacuum) so that atmospheric pressure forces the liquid through the filter. As the drum revolves, cake builds up on the surface of the cloth-covered drum. When it reaches a preset thickness, it is shaved off by a fixed blade, called a doctor blade, against which the drum rotates. Continuous filters may be operated for relatively long periods, but eventually the filter cloth becomes clogged ("blinded") by very fine particles, and the filter must be shut down until the cloth is cleaned or replaced. In addition to the continuous drum filter, there are many other kinds of machines, including travelling belt filters and rotating disk filters.

Another way of separating solids from liquids is by using a *centrifuge*. Centrifugation is like sedimentation, except that the force of gravity is replaced by *centrifugal force* generated by spinning the liquid at high speed. In the centrifuge, solids collect at the wall of the spinning container, and clear liquid is removed from the center. When sufficient solids have been collected, the centrifuge must be stopped and the solids scraped out. (Some centrifuges can be scraped automatically.)

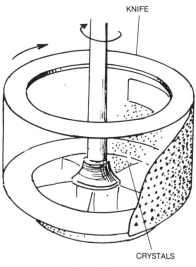

KNIFE

CRYSTALS

Centrifuge

Centrifuges can also be used for separating liquids that do not dissolve in each other, such as oil and water. Such centrifuges can be made to operate continuously, with the denser liquid being removed near the wall of the centrifuge and the other from the center. Cream is separated from milk in centrifuges of this type. In addition to separating liquids from each other—called liquid/liquid separations—and liquids from solids—called liquid/solid separations—there is the problem of separating liquids, solids, and gases from other gases—liquid/gas, solid/gas, and gas/gas separations. All of these are common in air pollution control applications.

One of the simplest devices for separating fairly large particles of solid or liquid from air or any other gas is the *cyclone separator*. Air passing through a cyclone is caused to swirl and the resulting centrifugal force drives the particles to the wall where they fall to the bottom and into a container. Clean gas exits from the center at the top.

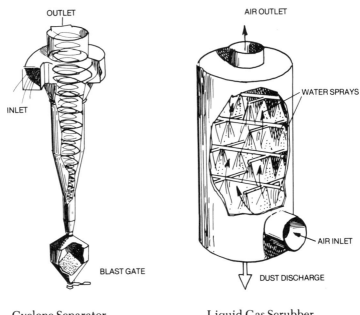

Cyclone Separator   Liquid Gas Scrubber

Where particles too fine for a cyclone separator need to be removed, the most common solution is the *gas scrubber*. Droplets of water are passed through the gas being scrubbed; they attach to the particles and carry them out of the scrubber. Scrubbers can also be used to remove one gas from another if one of the gases is soluble in water or some other solvent. Scrubbers are often used to remove sulfur dioxide or ammonia from air before it is released from a plant.

Discussion

1. What are some materials for which a grizzly might be used?

2. Are grizzlies used for large or small particles?

3. Are sieves used for large or small particles?

4. Of what is a sieve made?

5. What is a screen? How is it used?

6. How is a sieve analysis carried out?

7. What is the reason for using standard sieves?

8. What is the force that causes sedimentation?

9. Are dry solids discharged from a thickener?

10. Describe the process of filtration.

11. Why is filteraid used as a precoat?

12. What is the most common device used for batch filtration?

13. Where is filter cloth used in a plate-and-frame filter press?

14. In what part of a plate-and-frame filter press does cake accumulate?

15. What is the difference between filtrate and filter cake?

16. What must be done when a filter press is full of cake?

17. Why is this process expensive?

18. What are two advantages of filter presses?

19. Why are continuous filters expensive?

20. What is the most common type of continuous filter?

21. What is the name of the blade that shaves cake off a rotating drum filter?

22. How does a centrifuge work?

23. What kinds of materials can centrifuges separate?

24. For what reason might a chemical engineer want to separate gases or solids from air?

25. Describe a cyclone separator.

26. How does a gas scrubber work?

Review

A. Make a list of the various kinds of things that might be classified according to size by using sieves. Do not forget agricultural products.

B. Used lubricating oil from automobile engines contains fine particles of metal and other materials. Discuss how they might be removed from the oil, leaving it clean.

C. Match the terms in the left column with the proper short definition in the right column. Only one definition is appropriate for each term.

1. Sedimentation	_____ Separating solids according to size
2. Filtration	_____ A kind of sieve
3. Size classification	_____ Carries out sedimentation continuously
4. Filtrate	_____ A special finely woven cloth
5. Continuous filter	_____ Clear liquid from a filter

6. Cyclone separator   _____ Filters for a long time without stopping

7. Centrifuge   _____ Separates solid from gas using a water spray

8. Filter press   _____ Swirls a gas to separate solids from it

9. Grizzly   _____ Particles settle by gravity

10. Gas scrubber   _____ Liquid is forced through a cloth

11. Screen   _____ Parallel bars for size classification

12. Thickener   _____ A batch filter

13. Filter cake   _____ Uses centrifugal force to separate materials

14. Filter cloth   _____ Solids removed during filtration

UNIT TEN
CHALLENGES FACING
CHEMICAL ENGINEERING

Special Terms

Coal Liquefaction: A process by which a liquid fuel can be made from coal.

Coal Gasification: A process by which a fuel gas can be made from coal.

Substitute Natural Gas (SNG): A fuel gas, similar in heating value to natural gas, that can be made from coal or other materials.

Coal Conversion: Making a liquid or gaseous fuel from coal.

Oil Shale: A mineral that can be processed to make a petroleum-like product; the product is called *shale oil*.

Zero Emission: A concept under which a plant would not emit any polluting substances whatsoever. No zero-emission plant presently exists.

Biomedical Engineering: The science that uses engineering principles to help cure or alleviate disease.

Artificial Kidney: A device perfected by biomedical engineers that can keep people alive after their own kidneys have stopped working.

Technological Obsolescence: The process by which technical knowledge becomes outdated and less useful with the passage of time.

Continuing Education: Learning that is carried out from the time one leaves college to the end of one's career.

Computer Control: A way in which a computer is used to help run a process plant.

Mathematical Modeling: A way of expressing in mathematical terms what goes on in a chemcial process.

Computer Simulation: The use of a mathematical model in a computer so that an engineer can discover the effect of changing variables without actual experimentation in a plant or pilot plant.

Vocabulary Practice

1. What is *coal liquefaction*?

2. What is *coal gasification*?

3. What is *substitute natural gas*? How is the term abbreviated?

4. What is meant by *coal conversion*?

5. What is *oil shale*? What is the product made from oil shale?

6. What is the concept of *zero emission*? Do such plants exist?

7. What is *biomedical engineering*?

8. What is an *artificial kidney* used for?

9. What is *technological obsolescence*?

10. What is *continuing education*?

11. Define *computer control*.

12. What is done in *mathematical modeling*?

13. What is *computer simulation* used for?

Challenges Facing Chemical Engineering

There are serious problems facing the world today. Many of them challenge chemical engineers because they are the ones who may be able to find the solutions.

Perhaps our chief worldwide problem is a shortage of energy. For the past 50 years, most of the world's energy has come from petroleum and natural gas. But much of the existing supply has already been used up. If we continue to use these fuels at our constantly increasing rate, reserves will run out in perhaps 50 to 75 years.

So we must look for other sources and forms of energy. There is still a lot of coal left in the ground and it can be substituted for petroleum in important ways—such as for generating electricity. But nobody knows how to run an automobile or an airplane on coal. One way to help solve the problem of dwindling petroleum reserves is to make some sort of liquid fuel out of coal—a liquid that could be used to operate automobiles and airplanes. Doing this is called *coal liquefaction* and chemcial engineers have developed a number of processes for it. Unfortunately, the cost of fuel made by coal liquefaction is now about twice as high as for that made from petroleum. But the price of petroleum continues to rise and chemical engineers continue to find cheaper ways for coal liquefaction. Eventually, one of these processes will become economical.

It is also possible to make a substitute for natural gas from coal. This is called *coal gasification*, and the gas formed is known as *substitute natural gas* (SNG). Since similar processes are used for coal liquefaction and coal gasification, the two are sometimes considered together and called *coal conversion.*

Another substitute for petroleum can be made from a mineral called *oil shale.* This mineral can be processed by heat to yield a product very much like petroleum, called *shale oil.* There are two problems that chemical engineers are trying to work out: all the present processes require large amounts of water, and oil shale is found in very dry country; and oil shale expands when it is processed, so that there is a large amount of waste to dispose of.

Another pressing problem of the modern world is pollution. Many of the ways for cleaning air and water to prevent pollution are actually chemical processes and chemical engineers have been leaders in their development. One of the difficulties with pollution control is that many of the solutions to these problems create new problems of their own. The gas scrubber that removes particles from air by spraying droplets of water through it cleans the air very effectively, but it also leaves the engineer with the problem of disposing of a lot of dirty water. Indeed, a gas scrubber has been jokingly described as a machine that turns an air pollution problem into a water pollution problem. The usual solution is to remove the solids from the water by sedimentation and then to bury the solid material.

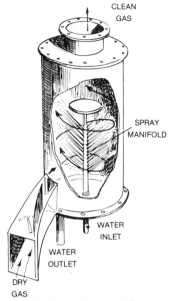

Cyclonic Gas Scrubber

The perfectly nonpolluting plant would be one that discharges no pollutants at all; this is known as *zero emission*. No process plant in the world yet runs on a zero emission basis, but chemical engineers have accepted this challenge.

In the past, human disease was considered a problem for the medical profession to solve alone. But the human body carries out many chemical reactions. In fact, the body has been described as a small, very complicated chemical process plant. Sometimes, when parts of the body break down, it is possible to use chemical engineering principles to make substitute parts. This is known as *biomedical engineering*. Until recently, when a person's kidneys stopped working, there was no way to continue the process of removing waste products from the blood, so the person died. Now these wastes can be removed by passing the blood through a machine that works on the same principles as the kidney—an *artificial kidney*. Such machines already keep thousands of people alive after their own kidneys fail. Another device developed by biomedical engineers is the heart/lung machine. It can take over the function of both the heart and lungs, allowing a heart to be stopped for surgical repair for as long as an hour.

The science of chemical engineering is growing in exciting di-

Courtesy National Institutes of Health
This heart/lung machine can take over the function of both the heart and lungs, allowing the heart to be stopped for surgical repair.

rections. New discoveries are constantly made and new processes invented from them and for them. Whatever a chemical engineer learns in college is soon replaced or modified with newer knowledge. This process is called *technological obsolescence* and simply means that an engineer's knowledge has become outdated. To prevent technological obsolescence, engineers have to keep learning new things. They can do this by reading current books and the technical magazines that serve their profession. Also, they can go back to school from time to time and take courses designed to keep their knowledge up to date. All this is known as *continuing education.*

One of the important ways in which chemical engineering has changed in the past 20 years has to do with computers. Control instruments can be tied in with a computer and the computer can thereby help run the plant. This is called *computer control* and is becoming more and more common. Computers can also be used to perform some of the design calculations that formerly had to be done by hand; heat exchangers can now be designed by computers.

Computer control of chemical plants is increasingly widespread.

The day when chemical processes and plants will be directed and controlled by computers is already in sight.

In addition, it has become possible to express what is going on in a process in the form of mathematical equations; this is called *mathematical modeling.* If a mathematical model is fed into a computer, it is possible to find out what will happen if some variable is changed—let us say, what will happen if the process is carried out at a higher temperature. Doing this is known as *computer simulation* of a process. Eventually it may be possible to use computer simulation as a substitute for pilot plants, but much more must be learned about chemical processes before this will become standard.

The future poses serious new problems—many of them about the supply and use of energy. It also raises new expectations—many of them about living conditions. Using their exciting new tools, chemical engineers will be able to meet these challenges. How well they do this will determine, in countless ways, the quality of life as we will know it.

Discussion

1. Why is there a shortage of energy in the world today?

2. What available fuel might be substituted for petroleum?

3. What is the process by which a liquid fuel is made out of coal?

4. What is the name of the process by which a fuel gas is made from coal?

5. What does SNG stand for?

6. What material is shale oil made from? Name two problems in making it.

7. Why are chemical engineers leading the development of pollution control processes?

8. What pollution problem does a gas scrubber solve? What problem does it cause?

9. What would you call a plant that produced no pollution at all?

10. What is the science of biomedical engineering about? Name some of its innovations.

11. What is outdated engineering knowledge called?

12. What is a way of preventing this problem?

13. Name three ways in which computers are used in chemical process plants.

Review

A. Which of the new developments described in this unit interests you the most? Would you like to work on any of them? If you would like to, explain why. If not, explain what kind of a career you would prefer to follow.

B. Based on your own knowledge and that given in this book, explain how pollution is controlled in either air or water.

C. Complete the following sentences with the proper word or phrase.

1. The science that uses engineering principles to solve medical problems is called _____.

2. The purpose of coal liquefaction is to make a _____ from coal.

3. When a chemcial engineer reads engineering magazines and books, it is part of the engineer's _____.

4. A gas scrubber turns an _____ pollution problem into a _____ pollution problem.

5. The letters SNG stand for _____.

6. The _____ machine has made lengthy heart surgery possible.

7. Shale oil is made from _____.

8. When an engineer expresses a process in the form of mathematical equations, he is doing _____.

9. _____ is the name of the process by which an engineer's knowledge gets outdated.

10. An engineer uses _____ to tell what will happen in a process by using a mathematical model.

11. Calculations that were formerly done by hand can now be done more quickly and accurately by using a _____.

12. _____ is the name of the process by which a computer is used to help run a chemical process plant.

13. The main function of the _____ is to remove waste products from the blood. If it fails, a person can be kept alive by using an _____.

14. One of the problems in making oil from oil shale is the absence of _____ in the area where oil shale is found.

INDEX OF SPECIAL TERMS